Machining of Light Alloys

Aluminum, Titanium, and Magnesium

Manufacturing Design and Technology Series
Series Editor: J. Paulo Davim

This *Manufacturing Design and Technology Series* series will publish high quality references and advanced textbooks in the broad area of manufacturing design and technology, with a special focus on sustainability in manufacturing. Books in the series should find a balance between academic research and industrial application. This series targets academics and practicing engineers working on topics in materials science, mechanical engineering, industrial engineering, systems engineering, and environmental engineering as related to manufacturing systems, as well as professions in manufacturing design.

Drills
Science and Technology of Advanced Operations
Viktor P. Astakhov

Technological Challenges and Management
Matching Human and Business Needs
Edited by Carolina Machado and J. Paulo Davim

Advanced Machining Processes
Innovative Modeling Techniques
Edited by Angelos P. Markopoulos and J. Paulo Davim

Management and Technological Challenges in the Digital Age
Edited by Pedro Novo Melo and Carolina Machado

For more information about this series, please visit: https://www.crcpress.com/Manufacturing-Design-and-Technology/book-series/CRCMANDESTEC

Machining of Light Alloys
Aluminum, Titanium, and Magnesium

Edited by
Diego Carou
J. Paulo Davim

CRC Press
Taylor & Francis Group
Boca Raton London New York

CRC Press is an imprint of the
Taylor & Francis Group, an **informa** business

CRC Press
Taylor & Francis Group
6000 Broken Sound Parkway NW, Suite 300
Boca Raton, FL 33487-2742

First issued in paperback 2020

© 2019 by Taylor & Francis Group, LLC
CRC Press is an imprint of Taylor & Francis Group, an Informa business

No claim to original U.S. Government works

ISBN-13: 978-1-138-74418-9 (hbk)
ISBN-13: 978-0-367-78099-9 (pbk)

Visit the Taylor & Francis Web site at
http://www.taylorandfrancis.com

and the CRC Press Web site at
http://www.crcpress.com

Contents

Preface

The use of light alloys is being increased in sectors such as aeronautics, automotive, and medicine. One of the drivers that make light alloys important materials in engineering applications is their relatively low density. Particularly, the development of these alloys is highly related to the needs of the transportation sector to diminish fuel consumption and the emissions to the atmosphere. But, in addition, strong efforts in the development of special designed alloys have been carried out in the last decades to meet new requirements in high demanding applications.

Main light alloys include aluminum, magnesium, and titanium. Despite their similar characteristics when considering density, the machinability of these alloys is different depending on the considered alloy. The special characteristics of these alloys challenge the knowledge of researchers to find the best solutions for machining them in the current framework guided by new paradigms such as Industry 4.0 and sustainable manufacturing.

The book gathers a collection of experimental and review studies around the machining of light alloys. The work is arranged in eight chapters that review in a comprehensive way the current knowledge on the machining of these materials, focusing on their main challenges and opportunities to improve their applications. Chapter 1 presents a revision of the main literature relevant to the machining of light alloys. Chapter 2 deals with a critical topic in the machining of aluminium alloys, particularly in dry machining, that is, the formation of built-up edge. Chapter 3 reviews one of the most important processes in machining, that is, drilling, in some of its variants. Chapter 4 presents a comprehensive experimental study from design to machining of hybrid parts of magnesium base material. Chapter 5 covers a nonconventional machining process (laser) for magnesium alloys. Chapter 6 focuses on sensor monitoring for a difficult-to-cut material, titanium. Chapter 7 studies the use of new techniques for cooling/lubricating, specifically cryogenic machining, in the machining of titanium alloys. And, finally, Chapter 8 analyzes drilling of light alloys and carbon-fiber reinforced plastic hybrid stacks for aerospace applications.

The editors would like to thank the authors for their work during the production of the book and valuable contributions.

Diego Carou
J. Paulo Davim
Aveiro, Portugal

Editors

Diego Carou received his PhD degree in industrial engineering from the National University of Distance Education (UNED) in 2013. He has international postdoctoral experience in manufacturing process at several european universities. He currently works as Assistant Professor at the University of Jaén. He has published more than 30 articles in journals and conferences, and book chapters. He also serves as reviewer for several international journals.

J. Paulo Davim received his PhD in mechanical engineering in 1997, MSc degree in mechanical engineering (materials and manufacturing processes) in 1991, and mechanical engineering degree (5 years) in 1986, from the University of Porto (FEUP); the Aggregate title from the University of Coimbra in 2005; and a DSc from London Metropolitan University in 2013. He is Eur Ing and Senior Chartered Engineer by the Portuguese Institution of Engineers, with an MBA and specialist title in Engineering and Industrial Management. Currently, he is professor at the Department of Mechanical Engineering of the University of Aveiro. He has more than 30 years of teaching and research experience in manufacturing, materials, and mechanical engineering with special emphasis in machining and tribology. Recently, he has also had an interest in management/industrial engineering and higher education for sustainability/engineering education. He has received several scientific awards. He has worked as an evaluator of projects for international research agencies as well as examiner of PhD thesis for many universities. He is the editor in chief of several international journals, guest editor of journals, book editor, book series editor, and scientific advisory for many international journals and conferences. Presently, he is an editorial board member of 25 international journals and acts as reviewer for more than 80 prestigious Web of Science journals. In addition, he has also published as editor (and coeditor) more than 100 books and as author (and coauthor) more than 10 books, 80 book chapters, and 400 articles in journals and conferences (more than 200 articles in journals).

Contributors

Eva María Rubio Alvir
Department of Manufacturing
 Engineering, Industrial
 Engineering School
Universidad Nacional de
 Educación a Distancia (UNED)
Madrid, Spain

Samir Atlati
CNRS, Arts et Métiers ParisTech,
 LEM3
Université de Lorraine
Metz, France

Lincoln Cardoso Brandão
Department of Mechanical
 Engineering
Federal University of São João
 del Rei
São João del Rei, Brazil

Stefania Bruschi
Department of Industrial
 Engineering
University of Padova
Padova, Italy

Alessandra Caggiano
Fraunhofer Joint Laboratory
 of Excellence on Advanced
 Production Technology
Fh-J_LEAPT UniNaples
Department of Industrial
 Engineering
University of Naples Federico II
Naples, Italy

Diego Carou
Department of Mechanical and
 Mining Engineering
Universidad de Jaén
Jaén, Spain

Mohd Danish
Mechanical Engineering
 Department
Universiti Teknologi Petronas
Seri Iskandar, Perak, Malaysia

J. Paulo Davim
Department of Mechanical
 Engineering
University of Aveiro
Aveiro, Portugal

Andrea Ghiotti
Department of Industrial
  Engineering
University of Padova
Padova, Italy

Turnad Lenggo Ginta
Mechanical Engineering
  Department
Universiti Teknologi Petronas
Seri Iskandar, Perak, Malaysia

Yingchun Guan
National Engineering Laboratory
  of Additive Manufacturing for
  Large Metallic Components
Beihang University
Beijing, China

Badis Haddag
CNRS, Arts et Métiers ParisTech,
  LEM3
Université de Lorraine
Metz, France

Carlos Henrique Lauro
Department of Mechanical
  Engineering
University of Aveiro
Aveiro, Portugal

and

Department of Mechanical
  Engineering
Federal University of São João
  del Rei
São João del Rei, Brazil

Jia Li
National Engineering Laboratory
  of Additive Manufacturing for
  Large Metallic Components
Beihang University
Beijing, China

Abdelhadi Moufki
CNRS, Arts et Métiers ParisTech,
  LEM3
Université de Lorraine
Metz, France

Luigi Nele
Department of Chemical,
  Materials, and Industrial
  Production Engineering
University of Naples Federico II
Naples, Italy

Mohammed Nouari
CNRS, Arts et Métiers ParisTech,
  LEM3
Université de Lorraine
Metz, France

Ahmad Majdi Abdul Rani
Mechanical Engineering
  Department
Universiti Teknologi Petronas
Seri Iskandar, Perak, Malaysia

José Manuel Sáenz de Pipaón
Department of Manufacturing
  Engineering, Industrial
  Engineering School
Universidad Nacional de
  Educación a Distancia (UNED)
Madrid, Spain

Roberto Teti
Fraunhofer Joint Laboratory
    of Excellence on Advanced
    Production Technology
Fh-J_LEAPT UniNaples
Department of Chemical,
    Materials, and Industrial
    Production Engineering
University of Naples Federico II
Naples, Italy

Xiangjun Tian
National Engineering Laboratory
    of Additive Manufacturing for
    Large Metallic Components
Beihang University
Beijing, China

José Luis Valencia
Department of Statistics and
    Operation Research III, Faculty
    of Statistical Studies
Complutense University of Madrid
    (UCM)
Madrid, Spain

María Villeta
Department of Statistics and
    Operation Research III, Faculty
    of Statistical Studies
Complutense University of Madrid
    (UCM)
Madrid, Spain

Muhammad Yasir
Mechanical Engineering
    Department
Universiti Teknologi Petronas
Seri Iskandar, Perak, Malaysia

Shuquan Zhang
National Engineering Laboratory
    of Additive Manufacturing for
    Large Metallic Components
Beihang University
Beijing, China

Hongyu Zheng
Singapore Institute of
    Manufacturing Technology
Singapore, Singapore

and

School of Mechanical Engineering
Shandong University of
    Technology
Zibo, China

chapter one

Light alloys and their machinability

A review

Mohd Danish, Turnad Lenggo Ginta, Muhammad Yasir, and Ahmad Majdi Abdul Rani

Contents

1.1   Introduction

The use of light alloys or materials having low specific weight is critical in manufactured parts used in aerospace, automobiles, electronics, sports, and biomedical applications. Generally, the term "light alloys" is used for alloys having base material aluminum (Al), magnesium (Mg), and titanium (Ti) (Polmear 2005). The density of the part becomes a critical factor when it comes to automobiles and mainly in aerospace. This is due to the fact that weight has a direct relation with the economy of the aerospace and automobile industries. A lighter vehicle will require less power, and a less amount of fuel will be consumed. In case of the space shuttle, mass is more critical as gravitational pull will be more powerful toward the greater mass. Light alloys have gained the attention of researchers around the globe because of not only their low density but also their other attractive properties, like the good machinability of magnesium, electrical and thermal conductivity of aluminum, and high corrosion resistance of titanium.

Machining is widely used for manufacturing products used in various industries like aviation, automobiles, and biomedical. There is not a fixed definition of machinability of materials, but still, it can be defined as ease of the material to be machined to have the required shape and size. It can also be defined from the power consumption perspective and the morphology of the chips (Elgallad et al. 2010). However, machining is not easy when it comes to light alloys. This chapter will review the properties and the machinability of light alloys focusing mainly on aluminum-, magnesium-, and titanium-based alloys. A special section for cryogenic machining of light alloys is also presented in this chapter.

1.2   Properties and applications

All alloys have different physical, electrical, and mechanical properties that make them suitable for some applications and unsuitable for some others. In this section, the properties and applications of aluminum, magnesium, and titanium alloys are presented.

1.2.1   Aluminum

Aluminum is the third most abundant metal in the earth's crust. It is found in the combination of oxygen and with other elements in nature. The structure of aluminum is face centered cubic and also has high ductility at room temperature (Hamade and Ismail 2005). Aluminum also has a relatively low melting point (about 660°C) compared to other engineering metals (Clayton 1987). The density of aluminum alloys is 30% to 35% of steels. Table 1.1 shows the mechanical properties of alloys of aluminum

Table 1.1 Mechanical properties of aluminum, magnesium, titanium, and steel

Mechanical properties		Aluminum alloys	AISI 4000 series steel	Magnesium alloy	Titanium alloys
Hardness	Vickers	28–79	121–578	30–600	290–411
	Brinell	29–89	3–700	59–100	304–480
Tensile strength	Ultimate (MPa)	90–295	450–1970	90–1070	825–1580
	Yield (MPa)	31–285	275–1860	21–460	759–1410
Elongation at break (%)		1–40	8–34	1–75	3–18
Modulus of elasticity (GPa)		68.9–70.0	196–213	38–120	105–125
Compressive yield strength (MPa)		0.552–4.60	1650–1800	21–448	860–1280
Poisson's ratio		0.330–0.350	0.270–0.300	0.270–0.350	0.310–0.342
Fatigue strength (MPa)		48.3–110	138–772	30–235	140–1160
Shear modulus (GPa)		0.04483–26.0	75–82	16.3–48.0	41.0–48.3

together with magnesium, titanium, and steels (Carou, Rubio, and Davim 2015).

Aluminum and its alloys have a low melting point. They are also highly resistant to corrosion (Davis 2007) and have high friction coefficient, excellent formability, and high magnetic neutrality (Hatch et al. 1984). The thermal properties of aluminum alloys are shown in Table 1.2 (Santos et al. 2016).

During the eighteenth century, Hall-Heroult electrolysis reduction was employed to produce aluminum alloys for the first time (Santos et al. 2016). Since then, aluminum alloys has seen a tremendous increase in production from 45,000 tons to more than 25 million tons nowadays. Since 1930, aluminum alloys have been used in constructing aircraft parts (Troeger and Starke 2000; Tan and Ogel 2007). Aluminum alloys present a high strength-to-weight ratio, and therefore, they have replaced steel and iron to manufacture parts (Miller et al. 2000; Hamade and Ismail 2005).

Table 1.2 Thermal properties of different light alloys

Thermal properties	Value		
	Al	Mg	Ti
Specific heat capacity (J/kg°C)	0.690–1.01	0.80–1.45	0.368–0.670
Thermal conductivity (W/mK)	1.48–255	44.3–159	6.10–10.9
Melting point (°C)	183–1350	330–650	1570–1700
Liquidus (°C)	184–660	585–650	1640 to 1700
Solidus (°C)	210–903	330–650	1300–1630

The demand for aluminum alloys in the automotive industries for manufacturing vehicles that are as light as possible with good mechanical properties makes them one of the most appropriate alloys to be used in this sector (Bishop et al. 2000; Ozcatalbas 2003). A significant amount of references can be cited for using aluminum alloys in automotive industries (Santos et al. 2016). The usage of aluminum per vehicle has seen an eight times increase over the last 50 years. Today, the amount of aluminum used per passenger vehicle is 180 kg. This figure is expected to reach 250 kg per passenger vehicle by 2025, which is about a 70% increase in the current value (Santos et al. 2016).

In the manufacturing of car wheels, panels, and structure, mainly the 6061 alloys are used (Demir and Gündüz 2009). Silicon carbide or aluminum oxide reinforced 6061 aluminum alloy is usually used for making pistons, brake discs, brake drum, and piston sleeves (Polini et al. 2003; Castro et al. 2008). 7075-T6 aluminum alloys are used in making fittings, gears, and shafts (Ng et al. 2006; Kannan and Kishawy 2008). 7050-T7451, 2024-T3, and 2014 aluminum alloys are used in manufacturing aircraft structures, the skin of aircraft, and rocket chambers, respectively (Alniak and Bedir 2003; Huda, Taib, and Zaharinie 2009; Tang et al. 2009). These are some major applications of aluminum alloys. In addition, aluminum alloys are also used in civil constructions, electrical devices, and in packaging industries. It is also used as the master mold to fabricate microfluidic devices (Basha et al. 2017; Yousuff et al. 2017).

1.2.2   *Magnesium*

Magnesium is one of the heavily utilized elements in the manufacturing industry. Magnesium is mostly found in the form of the compounds because of its high reactivity, and therefore, the production industry has a critical role in producing magnesium to be used in other engineering sectors like automobiles, aerospace, sports, and biomedical (Gray and Luan 2002). It is also one of the most abundant elements found on earth along with aluminum, calcium, and iron (McDonough and Sun 1995; Kıpouros and Sadoway 2001).

The nuclear industry and the military aircraft manufacturing industry are among the earliest references of utilization of magnesium, which falls in the period of World Wars I and II. However, there is a significant drop in its utilization post World War II, and it was in the last decade that the interest in magnesium has started to reemerge (Mordike and Ebert 2001). This can be established from the significant surge in the world primary production as well as an increase in the capacity of magnesium to be produced. The brightest spot in terms of the recent development in magnesium industry is in China. Among the most significant industries involving consumption of magnesium are the aeronautics, automotive, electronics, and medical (Scintilla and Tricarico 2013). This has served as a driving force for conducting further work toward a higher understanding of magnesium, as well as its utilization in the industrial processes, with expected results. One of the important areas that have emerged in this direction is the evaluation of traditional machining processes for their potential in delivering new magnesium-based parts or components and in repairing or maintaining magnesium-based components or products.

Magnesium's hexagonal structure restricts its movement around the basal planes, and hence, it is difficult to deform it at room temperature (Carou, Rubio, and Davim 2015). Typically, the temperature range for magnesium alloy formation is between 340°C and 510°C. For producing magnesium pieces, casting is the most suitable method for magnesium pieces due to the issue of cold forming associated with magnesium (Campbell 2006). Table 1.2 lists the details pertaining to the mechanical properties of magnesium alloys along with aluminum and titanium alloys (Carou, Rubio, and Davim 2015). The mechanical properties of magnesium alloy were almost equivalent to those of other structural materials having low densities. For some specific properties like resistance to dent, specific strength, and stiffness, magnesium alloys possess these in even higher levels as compared to other structural materials, along with its standout feature of low density, which makes it serve a great purpose in the competitiveness of magnesium as a structural material (Wang et al. 2008). Further to these specific properties, thermal properties are also important. During any machining process, temperature rises, which has an adverse effect on the quality of the machined component or the part and also affects the overall productivity of the process. Of all thermal properties, thermal conductivity is regarded as the most significant. It is due to the fact that a high value of thermal conductivity will help in evacuating the heat produced during any metal cutting process. Magnesium alloys also have a high value of thermal conductivity, which is beneficial during its machining process.

Magnesium alloys find use in a wide number of applications (Carou, Rubio, and Davim 2015). One such domain is their use in structural produces like die casting or wrought products (Carou, Rubio, and Davim 2015). Another area that has become popular of late is in the production of material matrix composites, which has shown potential to replace conventional materials for many structural applications like in transportation, aviation sector, defense, and sports (Wang et al. 2008; Vijaya et al. 2014). Magnesium can also be used as an alloying element. It is used as a reducing agent for aluminum, titanium, and zirconium (Westengen and Rashed 2016).

In the automotive industry, magnesium has been used for a long time for its contribution to the power consumption by means of significantly reducing overall weight. In a specific case, the utilization of magnesium in engine blocks in place of cast iron helped attain weight reduction higher than 50%, with 50 kg weight being reduced on this substitution (Tharumarajah and Koltun 2007). One of the first notable applications of magnesium was in Volkswagen Beetle, and since then, many popular brands such as Audi, BMW, Ford, Lexus, Toyota, and Porsche have all used magnesium in their products (Kulekci 2008; Carou, Rubio, and Davim 2015).

The improvement of solutions for the fuel usage of airships has, for all intents and purposes, vanished, and the utilization of magnesium is being constrained to motors and transmissions, particularly for helicopters (Polmear 2005; Carou, Rubio, and Davim 2015). The use of WE43 alloy for developing the transmissions for the MD500 and MD600 McDonnell Douglas helicopters can be highlighted (Froes, Eliezer, and Aghion 1998; Davies 2003). In addition, the utilization of QE alloy in the Anglo-French Jaguar contender in the form of nose wheel fork is another example (Charles, Crane, and Furness 1997). The QE22A alloy is additionally normally used for the development of gearboxes in flying machine structures (Kubiak et al. 2012). An essential field for magnesium combinations is their utilization in the motorsport. For example, in motorbikes, the wheels are made by die cast magnesium to reduce their weight and gain high speeds with minimum power requirement (Tharumarajah and Koltun 2007). As of late, the applications of magnesium composites in biomedical areas have likewise been discovered because of their low thickness, characteristic biocompatibility (Gray and Luan 2002), and adequate mechanical properties (Denkena and Lucas 2007). Magnesium and its alloys have been utilized as a part of various sorts of implants for both humans and animals amid the most recent 130 years (Witte 2015). For example, Mitsuishi et al. (2013) have reported the possible usage of AZ31 magnesium for making stents and the utilization of metal matrix composites with the AZ91D magnesium alloy matrix, and hydroxyapatite particles as fortifications have been explored in vitro by Witte et al. (Witte et al. 2007; Carou, Rubio, and Davim 2015).

1.2.3   *Titanium*

Titanium is the fourth richest structural metal on the earth's crust, which gained its importance in the manufacturing sector such as the aerospace, automotive, and biomedical industries. Titanium is considered as a good alternative to replace aluminum alloys and steel because of its exceptional properties like corrosion resistance, fatigue resistance, and a good strength-to-weight ratio, which are also well sustained at high temperature (Leyens and Peters 2003; Nourbakhsh et al. 2013; Veiga, Davim, and Loureiro 2013).

The structure of pure titanium is a hexagonal close-packed crystal structure, which is also called as "alpha" phase. At a temperature close to 900°C, this structure transforms to a cubic structure, which is known as "beta" phase (Peters and Leyens 2003). Mechanical properties play an important role in having good machinability of the alloy. For titanium alloys, chemical composition and microstructure are the key factors that determine their mechanical behavior. As we already know, the thermal properties of the material have a great influence on the machining process too. The main mechanical and thermal properties of titanium are shown in Tables 1.1 and 1.2 (Carou, Rubio, and Davim 2015; Matweb 2017).

Titanium and its alloy have vast application in almost every engineering sector. Titanium is used as an alloying element for steel to have reduced grain size. It is also used as a deoxidizing agent and to decrease carbon content in steel (Ezugwu and Wang 1997). Titanium alloys are available in the form of billet, bar, plate, sheet, extrusion, strip, hollows and wires, etc. (Trent 1977; Guimu et al. 2003; Peters and Leyens 2003). Moreover, titanium is a unique metal that is mostly used for alloying with metals like copper, iron, molybdenum, manganese, and other metals (Ezugwu and Wang 1997). Various products of titanium are being broadly utilized in aerospace, decorative, automotive, and as well as in medical applications. Apart from these, titanium alloys are also widely used in petroleum refineries, chemical processing, surgical implantation, pulp and paper industries, pollution control, nuclear waste storage, as well as electrochemical and marine applications. The demand of application of titanium and its alloys has seen a rapid rise. In 1958, Boeing 707 has 0.5% weight of titanium alloy, but in 1995, it was increased to 8.5% in Boeing 777 (Guimu et al. 2003). In the aircraft industry, where high fatigue is involved, titanium alloys are used for making spacers, bolts, springs, fasteners, etc. Structural stability is one of the important factors of jet engines, and in this sense, titanium alloys are used for the manufacture of flange rings, spacers, and bolts for these engines (Lei and Liu 2002; Carou et al. 2017). Titanium alloys have the ability to withstand high temperature, with permanent deformation. These alloys have a high corrosion resistance, which makes them suitable for applications in processing and chemical industries for the manufacturing of heat exchangers, pressure vessels, tanks, pumps, etc. In biomedical industries, titanium alloys have gained popularity for use in heart valve and dental implants due to their low density and high strength (Fraker et al. 1983).

1.3   Machinability of aluminum

Machining or metal cutting is related to the chip formation process as described by Trent (1977) in his classical book *Metal Cutting*. In the case of ductile materials such as aluminum, the tool–chip contact area is usually large during the cutting process. The thickness of the chip is also high, which gives rise to the cutting forces and thereby increases the required machining power. This high tool–chip contact area and high chip thickness further increase the heat generation. Usually, the morphology of the chips during machining of ductile material is long and stringy. All these factors, like the high generation of heat and long chips, lead to poor surface finish. However, in the case of aluminum alloy, the shear strength is relatively low, which makes their machining considerably easy even with the large tool–chip contact area. This section will present and discuss the main machining factors that directly or indirectly govern the machinability of aluminum alloy, like forces, cutting temperature, cutting tools, and tool wear, which will help in understanding the machining response of the alloy.

1.3.1   Cutting temperature

It is reported that the temperature rise during the machining of aluminum alloy does not produce any critical problem (Kelly and Cotterell 2002; Kishawy et al. 2005). However, at higher temperatures, the ductility of aluminum alloy can increase, which results in unwanted chip morphologies (Ozcatalbas 2003). Furthermore, the chemical reaction between the coating of the coated tool and aluminum can be initiated at high temperatures, which gives rise to diffusion between the two (Dwivedi, Sharma, and Rajan 2008). Moreover, due to the high mechanical strength, the presence of Si, SiC, and Al_2O_3, and high cutting speeds, temperatures can increase rapidly (Dwivedi, Sharma, and Rajan 2008; Zaghbani and Songmene 2009) due to high friction between the harder particles and the tool surface (Trent 1977). Generally, shear deformation rate increases with the increase in cutting speed, which results in higher cutting temperature (Braga et al. 2002). High feed rates also contribute to elevating the temperature, if the tool–chip contact area has not increased, because a higher tool–chip contact area can dissipate the heat at higher rates, as illustrated in Figure 1.1 (Zaghbani and Songmene 2009).

In the case of drilling of 2024-T351 aluminum alloys, Nouari et al. (2003) conducted finite element simulation of the process and investigated the impact of feed rate on the temperature at the tool–chip interface. They reported that with the increase in feed rate, the temperature at the tool–chip interface increased, which is shown in Figure 1.2. It can be due to the increase in material removal rate, which required high machining forces; variation in the thermal conductivity of the material at the different temperature; and chip morphology that usually occur during the drilling. Changes in the tool dimensions that can occur due to build-up edges (BUEs) or tool wear can also increase the temperature during the machining of aluminum alloys. Tool wear can give rise to the cutting forces, which can act as a new source of heat generation between the tool and machined material (Tang et al. 2009).

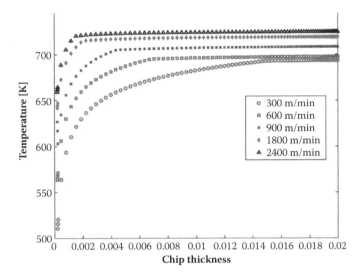

Figure 1.1 Influence of chip thickness and cutting speed on the temperature in the secondary shear zone for 7075-T6 aluminum in a half immersion up-milling. (Reprinted from *Journal of Materials Processing Technology*, 209, I. Zaghbani, V. Songmene, A force–temperature model including a constitutive law for dry high speed milling of aluminium alloys, 2532–2544, Copyright 2009, with permission from Elsevier.)

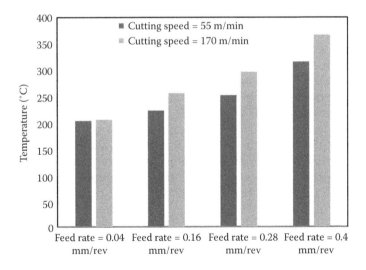

Figure 1.2 Influence of feed rate and cutting speed on the temperature in drilling of AA 2024 T35.1. (Data from Nouari, M., List, G., Girot, F., Coupard, D., *Wear*, 255, 1359–1368, 2003.)

1.3.2   Cutting forces

Cutting forces are generally lower in case of machining of aluminum and its alloys due to lower mechanical strength when compared to other structural materials. Among aluminum alloys, the cutting forces depend on their chemical composition, which also governs their other properties. The range of cutting forces in the aluminum alloy is not so wide (Campatelli and Scippa 2012; Salguero et al. 2013; Zhang et al. 2015). Furthermore, any process that increases the strength and hardness of the aluminum alloy, such as adding other elements in the alloy or mechanical treatment, may decrease the machining forces due to the reduction in the tool–chip contact area (Fang and Wu 2005; Chambers 1996). Normally, during the machining of any aluminum alloy, machining forces tend to decrease with an increase in the cutting speed irrespective of the strength of the material (Fuh and Chang 1997; Demir and Gündüz 2009; Manna and Bhattacharyya 2002). It is due to the fact that the shear stresses associated with the primary as well as secondary shear zones tend to reduce with the rise in cutting temperature, which occurs due to high cutting speeds (Kilic and Raman 2007; Zaghbani and Songmene 2009). However, machining forces reduce with an increase in cutting speeds due to thermal softening, and machining forces may rise during high-speed cutting operations due to high deformation rates (Hamade and Ismail 2005; Larbi, Djebali, and Bilek 2015). Yousefi and Ichida (2000) have also reported that a rise in cutting speeds tends to increase the cutting (when the cutting speed was greater than 100 m/s) during machining of aluminum magnesium alloy with tungsten carbide tool, as shown in Figure 1.3.

Furthermore, shear plane areas become larger when machining is done at higher feed rates or depth of cut, which may also result in higher machining forces (Fuh and Chang 1997; Manna and Bhattacharyya 2002; Ng et al. 2006). The stresses in the primary shear zone are around 30% higher than the stresses in the secondary shear zone, which is due to the lower temperature occurrence in the former (Zaghbani and Songmene 2009). During the machining of aluminum alloys, changes in the tool dimensions that can occur due to the BUEs or by wear have a strong effect on the machining forces predominantly if the dimensional change has affected the nose radius and/or rake angle of the tool (Saglam, Unsacar, and Yaldiz 2006; Gómez-Parra et al. 2013).

Tang et al. (2009) have reported the occurrence of excessive machining forces during the milling of 7050-T7451 aluminum alloy due to flank wear (Figure 1.4), which is due to the increase in the contact area of the tool work material. Machining forces can be reduced by polishing the cutting tool by chemical vapor deposition technique during the machining of aluminum alloys, which decreases the flank wear as well as enhances the surface finish (Arumugam, Malshe, and Batzer 2006; Castro et al. 2008).

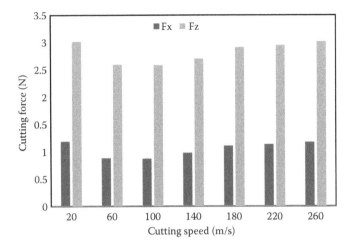

Figure 1.3 Variation of cutting forces with cutting speed. (Data from Yousefi, R., Ichida, Y., *Precis. Eng.*, 24, 371–376, 2000.)

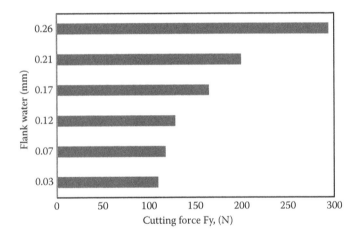

Figure 1.4 Effect of flank wear on cutting forces in milling of aluminum alloy. (Data from Tang, Z.T., Liu, Z.Q., Pan, Y.Z., Wan, Y., Ai, X., *J. Mater. Process. Technol.*, 209, 4502–4508, 2009.)

1.3.3 Cutting tools

A wide range of cutting tools is used when machining aluminum, such as high-speed steels, due to less reactivity with aluminum straight grade (K) of cemented carbides, which also significantly increases the surface finish, and tools that are based on diamond. Furthermore, diamond-based cutting tools also decrease the adherent layer of the tool edge in the chip flow direction (Pascual Serrano, Vera Pasamontes, and Girón Moreno 2016). Generally, for machining processes like milling, turning, drilling, and boring, use of K10 grade tool is suggested in the case of silicon aluminum alloys. Toropov, Ko, and Kim (2005) have tried a K10 grade tool having different rake angles ranging from −5° to 20° in turning Al6061-T6 alloy at a cutting speed of 800 m/min. They reported that the burr was smaller for the tools having higher rake angles (10° and 20°) than for the tools with smaller rake angles. Irrespective of the tool material machining of aluminum alloys containing hard particles around 15%, smaller cutting speed (20 to 450 m/min) was recommended with smaller tool rake angles (0° to 7°) (Trent 1977). Aluminum alloys having abrasive particles are suggested to be machined using K01 grade tools, and to cope with a sudden temperature rise in intermittent or interrupted machining, K20 grade tools are recommended (Rawangwong et al. 2013). The K20 grade tools were also widely used for machining of other structural materials (Liew, Hutchings, and Williams 1998).

While machining aluminum alloy having 12% of silicon carbide particles, the maximum cutting speed obtained was 225 m/min using an uncoated tungsten carbide cutting tool (Manna and Bhattacharyya 2002). While drilling aluminum alloys, high-helix-angle drills (40°) are used for relatively lower-strength alloys, whereas smaller-helix-angle drills (30°) were recommended for hard alloys (Hamade and Ismail 2005).

If the aluminum alloy contains very hard particles like ceramic or silicon, polycrystalline diamond (PCD) tools were recommended as they can withstand the wear mechanism more efficiently (Castro et al. 2008; Lane et al. 2010). PCD can effectively cut harder particles like ceramic particle during the machining of aluminum alloys than cemented carbide tools can (Mackerle 1999). Furthermore, the thermal conductivity of PCD tools is four times greater than that of cemented carbide cutting tools, and therefore, heat generation is quite lower when machining of aluminum alloy was done using PCD (Coelho et al. 1995). When the shape of the tool required is complex and the tools made from the solid diamond are difficult to manufacture, tools having diamond coating are recommended. This is due to their high thermal conductivity, stability in hardness even at high temperatures, less reactivity, and low friction coefficient, which enhances the machinability of the aluminum alloys (Paulo Davim and Monteiro Baptista 2000; Polini et al. 2003; Dasch et al. 2006).

1.3.4 Tool wear

Different types of wear mechanism can occur during the machining of aluminum and its alloys depending upon the machining conditions, work material, and the tool used. The most common wear that occurs during the machining process is flank wear, which is a result of adhesive and/or abrasive wear mechanisms (Nouari et al. 2003; Kannan and Kishawy 2008). Adhesive wear mechanism is generally initiated by the chemical reaction between the tool and work material, normally at higher cutting temperatures, due to which work material diffuses on the tool surface, and due to repeated cutting process, erosion of the defused material together with tool material occurs. Abrasive wear mechanism generally occurs due to the presence of hard particles in the metal matrix. These particles may erode the cutting tool. Therefore, the machining of the alloy containing hard particles like silicon carbide or silicon or ceramic particles to about 15% by volume of the alloy at high speeds may accelerate the wear rate (Chambers 1996).

It is reported that the wear of cutting tools is higher for machining of aluminum alloys containing hard particles such as aluminum oxide, silicon carbide, and silicon (Kelly and Cotterell 2002). During machining of the aluminum alloy containing hard particles, the temperature can be abruptly high due to the intermittent cutting of these particles, which further leads to tool wear (Trent 1977). Normally, the wear of the tool occurs near the cutting edge. Generally, the hardness of particles like silicon carbides present in the aluminum matrix to about 20% is higher (1.5 times) as compared to that in cemented carbide cutting tools (Rawangwong et al. 2013). Higher flank wear was observed by Coelho et al. (1995) while drilling was performed on aluminum alloys having silicon and silicon carbide particles using PCD tools (Figure 1.5).

The size of the hard particles, as well as their proportion present in the aluminum matrix, increases the flank wear rate (Biermann and Heilmann

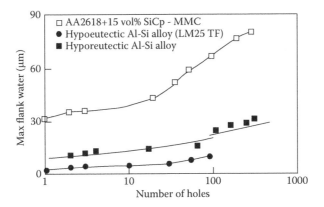

Figure 1.5 Flank wear evolution when drilling Al-Si alloys and MMC using PCD tipped drills. (Reprinted from *Journal of Machine Tools and Manufacture*, 32, R.T. Coelho, S. Yamada, D.K. Aspinwall, M.L.H. Wise, The application of polycrystalline diamond (PCD) tool materials when drilling and reaming aluminium based alloys including MMC, 761–774, Copyright 1995, with permission from Elsevier.)

2010; Beranoagirre and López De Lacalle 2011). The size of silicon particle and its distribution in the aluminum matrix can be controlled by the cooling rate (Ng et al. 2004). Aluminum and its alloys possess low melting points and therefore exhibit low machining temperatures, due to which wear rate is generally low and the deformation of the tool edges is also much less (Trent 1977).

In addition to flank wear, crater wear and notch wear were also observed while machining aluminum alloys having 12% wt. silicon particles using titanium carbide and tungsten carbide tools. These types of wear were reported in the study conducted by Liang, Vohra, and Thompson (2008), where the authors machined aluminum alloy having 18% wt. silicon using PCD tools. Ciftci, Turker, and Seker (2004) have also observed wear and notching of the tool while performing turning process on the 2014 aluminum alloy, where the tool material was cubic boron nitride (CBN) and having 0° rake angle. This behavior of the tool wear can be due to adhesion or abrasion, which may occur during the machining process (Andrewes, Feng, and Lau 2000). Smearing of work material has been also observed while machining aluminum alloys (Ng et al. 2006). This can be due to the increase in tool cutting edge due to flank wear during the machining process while using a carbide cutting tool. On the other hand, smearing of the work material was not observed while using PCD cutting tool irrespective of the flank wear.

1.4   Machinability of magnesium

Magnesium offers superior machinability as compared to the other structural materials. The properties considered for evaluating the machinability of magnesium alloys here include the morphology of the chips, cutting speeds, cutting forces, surface finish, and tool life (Polmear 2005; Shokrani, Dhokia, and Newman 2012). The morphology of the chips is governed by the composition of the alloy, temperature, and feed rate (ASM 1989). In the turning of magnesium, the influential parameter is tool geometry, tool rake angle in particular (Anilchandra and Surappa 2010). In addition to these advantages in terms of machinability, magnesium consumes about 8 to 10 times less power as compared to aluminum, stainless steel, mild steel, and cast iron (Polmear 2005).

In this section, the different aspects of machinability of magnesium and its alloys are presented in terms of temperature, cutting speed, tool wear and burr formation, and choices of the cutting tool.

1.4.1   Cutting temperature and ignition

The high specific heat and thermal conductivity of magnesium facilitate lower heat generation as the heat gets dissipated quickly (ASM 1989).

Apart from surface quality and tool wear, temperature is also important from a safety point of view. There is ignition risk when the temperature exceeds 405°C and temperature reaches around 3000°C when fire starts. Apart from maintaining safe temperature, it is important to keep particle size to less than 500 μm to keep it secure from any danger of explosion (Weinert et al. 2004; Danish et al. 2016, 2017).

Overall, the possibility of magnesium depends on a combination of factors, which include the alloy or material used, machining conditions, and the geometry of the tool (Zhao, Hou, and Zhu 2011). Generally, it is considered that a cutting speed lower than 300 m/min and feed rates in excess of 0.02 mm/rev significantly lower the risk of ignition that is associated with magnesium alloys (Destefani 1990).

Another important consideration is the use of cooling/lubricating systems, as there is also the ignition risk associated with it. In particular, water-based fluid poses an additional danger because magnesium can react with water and forms hydrogen, which can be highly flammable and explosive, which is also shown by the chemical reaction given by Equation 1.1 (Kulekci 2008).

$$Mg + H_2O \rightarrow MgO + H_2 \tag{1.1}$$

To counter this issue, mineral-based oils are recommended or dry machining is encouraged (ASM 1989; Machado and Wallbank 1997; Villeta et al. 2011). Also, there have been few studies where the feasibility of water was mentioned by choosing appropriate process parameters (Gariboldi 2003). Tomac and Tonnessen presented the suitability of having 5% mineral oil emulsion in water, which also helped prevent flank build-up (Tomac, Tonnessen, and Rasch 1991).

The other possible steps to prevent fire occurrence is by reducing the amount of chips from the working area to reduce the heat accumulation. In this direction, either the coolants that are normally used during machining processes or, in case of dry machining, cyclone-separator-based chip removal systems can be used. Placing the chips in a different container close to the machining zone can also be used (Weinert et al. 2004).

For additional safety of the operators, safety kits and instrument that can withstand high temperatures (fires/explosions) should be employed in the working area. Other basic protection systems include installation of alarm systems, placing fire extinguishers (preferably Class D type), and using dry sand containers (Rubio et al. 2012).

1.4.2   Cutting speeds

Magnesium provides the possibility of using higher cutting speeds when compared with the other structural metal and alloys. Information regarding the possible and safe cutting speeds for different magnesium alloys can be obtained from specialized works or in the documents that were provided by the tool manufacturers.

During face milling, cutting speeds from 4000 to 6000 m/s were found possible for drilling magnesium alloy having 0.6 mm feed per tooth by Byrne, Dornfeld, and Denkena (2003). They also reported that for drilling magnesium alloy when the feed rate was 0.8 mm/rev, the possible cutting speed was over 1000 m/min and for finishing operations, the possible cutting speed was above 1500 m/s.

1.4.3   Cutting tool

In making the choice for the appropriate tool, it must be noted that tool life has an adverse effect on the productivity correlated to the process and an optimized approach needs to be taken. Among the available choices in the literature is the PCD on account of its long life, good surface finish, and dimensional tolerance of the machined part (Dornfeld and Min 2010). Adhesion and abrasion because of reinforcing elements make machining a little difficult in case of magnesium metal matrix composites (MMCs) (Grzesik and Grzesik 2017). However, in the case of magnesium, the use of other types of tools is also possible because of its good machinability. Carbide tools have also been used, with good results (Tomac, Tonnessen, and Rasch 1991; Pu et al. 2012; Outeiro et al. 2013; Shen et al. 2017).

The design in the case of magnesium tools can be similar to the one adopted in the case of aluminum or steel machining. However, its low resistance and low heat capacity give the leverage of having products with a high-quality surface finish, greater relief angles, big chip spaces, lower number of tool cutters when doing milling operations, and smaller rake angles (ASM 1989).

1.4.4   Tool wear

Tool wear is generally much less in case of machining of magnesium alloys, especially in those without abrasive particles such as aluminum zinc magnesium alloys because they are not very adhesive to the tool surface and the absence of abrasive particle increases the tool life (Grzesik and Grzesik 2017). Under similar machining conditions, the life of the tool while performing dry machining of aluminum is five times lower than that of dry machining of magnesium (Dornfeld and Min 2010). Adhesion is more likely at higher cutting speeds due to the increase in temperature, which also reduces the surface finish. Normally, adhesion of the work material takes three different modes, which are BUEs, layer build-up, and build-up at flank face. In case of machining of magnesium and its alloys, the build-up layer at the flank face of the tool is most common

(Tomac, Tonnessen, and Rasch 1991). Also, moderate cutting speeds are recommended to avoid burrs, which is another phenomenon associated with machining (Pu et al. 2012; Danish, Ginta, and Wahjoedi 2016; Shen et al. 2017). The influencing factors in this method are the machining conditions, work material, and the dimensions of the tool.

1.5   Machinability of titanium

Titanium is a difficult-to-machine material because of its high chemical reactivity with tool materials, due to which welding occurs between the work material and the tool, which causes high surface roughness and blunt tools (Lei and Liu 2002; Ezugwu 2005). Furthermore, its low thermal conductivity (7.3 W/mK) is also one of the factors that make its machining difficult (Carou et al. 2017). The improvement in the machining of titanium alloys depends on the minimizing of the major problems associated with it, such as high cutting temperature, high cutting stress, chatter, and chip morphology, which are discussed in the following.

1.5.1   Cutting temperature

The machining of titanium alloys usually generates a high cutting temperature. When the temperature at the cutting edge rises, it wears out the tool rapidly. Approximately 80% of the generated heat is conducted in to the tool, and it cannot be removed with the fast flow of the chip and is transmitted in to the workpiece as the titanium has a low thermal conductivity (Ezugwu and Wang 1997). From the literature, it is evidenced that the cutting temperature has a great effect on the tool life and tool wear. Menezes et al. (2016) have reported an increase in temperature as well as increase in flank wear during milling of Ti6Al4V alloy, as illustrated by Figure 1.6.

Moreover, a high cutting temperature strongly affects surface integrity and chip formation mechanism and adds to the thermal deformation of the cutting tool, which is the main problem in the machining process (Abukhshim, Mativenga, and Sheikh 2006). During the cutting process, heat is generated in the primary shear zone, which flows in to the secondary zone. In the previous location, the workpiece experienced plastic deformation and high strain rates, which are in the order of 103 to 104 per second with different cutting speeds (Ng et al. 1999). The heat produced is more intense in machining of titanium alloys because it requires much more energy as compared to low-strength materials. Moreover, the conductivity of titanium is 15 W/m °C, which is much lower than that of steel. This makes the cutting temperature in the workpiece and the tool rise during machining (Wang and Rajurkar 2000). The overall temperature on the rake face of the tool is proportional to the cutting speed, thermal conductivity (K), specific cutting energy, the feed rate and uncut chip

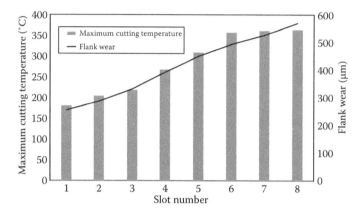

Figure 1.6 Variation of maximum cutting temperature and flank wear with different slot number when milling Ti6AL4V alloy. (Data from Menezes, J., Rubeo, M.A., Kiran, K., Honeycutt, A., Schmitz, T.L., *Proc. Manuf.*, 5, 427–441, 2016.)

thickness, the density (ρ), and the specific heat of the work piece (C). The product of K.ρ.C increases the temperature even at relatively low cutting speed and feed rate (Kitagawa, Kubo, and Maekawa 1997). During the cutting, the contact area of the tool with the workpiece is very small, causing highly concentrated stress and heat generation on the tool nose (Hong and Ding 2001). Ezugwu and Wang (1997) have reported that the cutting temperature was increased with an increase in cutting speed during the turning of titanium alloy Ti-6Al-4V, as shown in Figure 1.7.

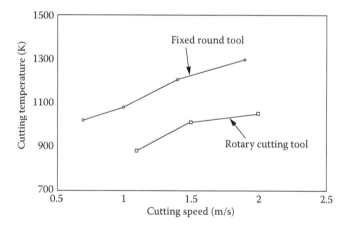

Figure 1.7 Temperature versus cutting speed in turning of Ti-6Al-4V. (Reprinted from *Journal of Materials Processing Technology*, 68, E.O. Ezugwu, Z.M. Wang, Titanium alloys and their machinability—a review, 262–274, Copyright 1997, with permission from Elsevier.)

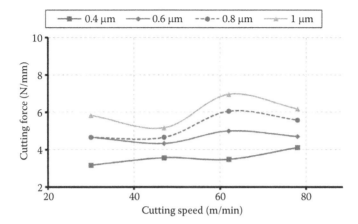

Figure 1.8 Cutting forces as a function of cutting speed. (Reprinted from *Journal of Materials Processing Technology*, 235, S.N.B. Oliaei, Y. Karpat, Investigating the influence of built-up edge on forces and surface roughness in micro scale orthogonal machining of titanium alloy Ti6Al4V, 28–40, Copyright 2016, with permission from Elsevier.)

1.5.2   Cutting forces

The cutting forces during machining of titanium alloys are lower than those obtained during machining of steels. At the initial stage of the cutting, the cutting forces are increased due to the change in contact length on the tool tip and the formation of BUE on the rake face. The increase in the average cutting force may be attributed to the rounding of the tool tip and BUE formation. Oliaei and Karpat (2016) reported an increase in thrust forces with an increase in cutting speed during orthogonal machining of the Ti6Al4V alloy, as shown in Figure 1.8. They mentioned that the increase in thrust forces was due to the formation of BUE on the tool face.

Stresses on the titanium alloys during machining are higher as compared to Nimonic 105 (nickel-based alloy) and steel Ck 53N. At elevated temperatures, titanium alloys show resistance to deformation, which reduces only above 800°C. During cutting, high stresses were applied on the cutting edge of the tool, which minimizes the contact surface and resulted in low plasticity of titanium alloys (Wang, Wong, and Rahman 2005).

1.5.3   Cutting tools

Cutting tools have great advancements in today's world. Various cutting tools such as ceramics, cubic-boron nitride, PCD, uncoated and coated carbides cutting tools, etc., are used nowadays according to the machining conditions and the work material. Hardness, wear resistance, and toughness are among the important characteristics of the tool that influence its performance (Haron et al. 2007). The hardness of the tool is also an important property. A harder tool is more stable against shock, and high cutting forces thus resist any permanent change in shape or geometry. But it also increases the brittleness of the tool, which increases the possibility of tool fracture. Therefore, it is usually recommended that the cutting tool should follow the following criteria:

a. The tool should have a high hot hardness in order to prevent the high stresses involved during machining. Tungsten carbide and PCD tools have a hot hardness value of 1800 HV and 10,000 HV, respectively (Shaw 2005).
b. Good thermal conductivity to resist thermal shock and thermal gradients. Tungsten carbide and PCD tools have 100 W/m °C and 560 W/m °C of thermal conductivities, respectively (Trent 1977).
c. Good chemical resistance against reaction with titanium.
d. They should have good toughness and resist fatigue to bear the chip segment formation process. Tungsten carbides have 9 MPam-1/2, while PCD tools have 8.8 MPam-1/2, respectively (Trent 1977).
e. High shear, compressive, and tensile strength.

For machining titanium at high cutting speeds, uncoated cemented carbide cutting tools are normally preferred (Jawaid, Che-Haron, and Abdullah 1999). During cutting, severe chipping and flaking of the cutting edge occur during milling of titanium alloys with carbide tools. These types of failures are due to the high thermomechanical and cyclic stresses, adhesion, and tool face wear of the workpiece material (Jawaid, Sharif, and Koksal 2000). Moreover, machining at a high cutting speed tends to increase the temperature at the cutting zone, especially at the edge of the tool, which results in excessive stresses at the tool nose, producing plastic deformation and tool wear (Che-Haron and Jawaid 2005). Sintered carbide tools are used in the machining of titanium alloys when the conventional speeds range from 30 to 100 m/min, resulting in low productivity (Su et al. 2006). The reactivity of titanium alloys is very high with the tool materials especially at a higher (about 500°C) cutting temperature. Titanium atoms diffuse into the carbide cutting tool, which leads to a chemical reaction with carbon and forms a layer of titanium carbide (TiC) at this temperature. This chemical interaction is insignificant between carbide and titanium alloys at low cutting speeds. At low cutting speeds, tool wear is basically caused by thermal fatigue, mechanical fatigue, and microfractures. Tool particles are plucked off from those microcracks, which are sandwiched between the tool and the workpiece, giving rise to abrasion wear.

Coated carbide is also used for the machining of titanium alloys in the speed range of 50–100 m/min, especially during the turning process (López De Lacalle et al. 2000). The coating is a thermal insulation for the tool and has a low coefficient of friction. This minimizes the cutting forces that result from the machining of titanium alloys. However, the coating is lost by delamination as the intermittent cutting forces come in to play during milling.

BCN and PCD tools are widely used in cutting of titanium alloys as the tool materials advance and they have more strength at high temperatures of 1100°C, where the carbide tools bear plastic deformation at this range of temperature (Ezugwu et al. 2005), while the titanium reacts with BCN tools at the temperature range of 800°C–850°C, which influences the tool wear to a great extent.

1.5.4   Tool wear

During machining of titanium alloys, the cutting tool faces intense thermal and mechanical shocks. As a result, high temperature and cutting stress generated near the cutting edge of the tool increase the tool wear rate and hence shortens the tool life (Ezugwu et al. 2005). Chipping, crater wear, flank wear, notching, and catastrophic failure are the main failure types while machining titanium alloys. During machining, the tool comes in contact with the workpiece, on which it exerts a strong compression force and the workpiece deforms plastically. Severe shear is generated in the primary zone between the tool and the workpiece, which starts the formation of the chips (Sridhar et al. 2003).

During machining, the temperature at the cutting edge rises and high stresses are induced, which results in chipping of the cutting edge of the tool. That chipped tool material flowing between the tool and the workpiece machined surface results in abrasion at some places. Adhesion and dissolution–diffusion are the wear mechanisms for the cutting tool when machining of Ti-6242S is done in a dry atmosphere, as reported by Haron et al. (2007), which is shown in Figure 1.9.

1.5.5   Chatter

During high surface finish machining of titanium alloys, chatter is one of the major problems that need to be addressed. Chatter during machining occurs because of the low modulus of elasticity of titanium alloys. Titanium alloys deflect twice as much as carbon when cutting pressure is applied to it. This creates a high spring-back behind the cutting edge, which results in premature flank wear, high cutting temperature, and vibration during machining (Ezugwu and Wang 1997). Dynamic cutting forces during machining of titanium alloys are also a cause of chattering,

Figure 1.9 Typical wear on a tool flank face (100 m/min, 0.15 mm/tooth, 20.4 min, VB3 0.3 mm). (Reprinted from *Journal of Materials Processing Technology*, 185, C.H.C. Haron, A. Ginting, H. Arshad, Performance of alloyed uncoated and CVD-coated carbide tools in dry milling of titanium alloy Ti-6242S, 77–82, Copyright 2007, with permission from Elsevier.)

which is 30% of the static forces (Antonialli, Diniz, and Pederiva 2010). These forces are due to the catastrophic or adiabatic thermoplastic shear process by means of which titanium chips are produced. Moreover, inappropriate machining parameters lead to chattering and considerably decrease the performance of the entire machining operation, which consists of the cutting tool, the machine tool fixture, and also the product (Abhang and Hameedullah 2010).

1.6   Cryogenic machining of light alloys

Nowadays, cryogenic machining is emerging as a prominent machining process for machining light alloys (Jawahir et al. 2016; Xu et al. 2016). The term "cryogenic" can be defined as working or processing the material at a cryogenic temperature, which is considered below –150°C (Liu, Chung, and Park 2014). In the machining process, cryogenic machining refers to machining the material at a very low temperature (below –150°C), which can be done by spraying or targeting the jet of liquefied gases such liquid nitrogen and liquid carbon dioxide at the machining zone (Dilip Jerold and Pradeep Kumar 2012; Danish et al. 2018). Cryogenic machining provides beneficial effects such as reducing the temperature at the cutting zone dramatically; less tool wear; providing a clean process by eliminating the slurry of the coolant and the chips, which is normally present when machining is done in the presence of cutting fluids; diminishing the harmful gases than can be produced due to the reaction between tool material and/or with work material due to the high temperature at high cutting speed machining; reducing the friction between the tool and work material; and also reducing machine vibrations due to the cushioning effect produced by the phase change of the cryogen (from liquid to gas)

(Hong, Ding, and Jeong 2001; Venugopal, Paul, and Chattopadhyay 2007; Dinesh et al. 2015; Sun et al. 2015b; Zhang et al. 2015; Jawahir et al. 2016; Danish et al. 2017). In one study, a dramatic increase in cutting tool life was observed during cryogenic machining as compared to dry machining, where the work material was Ti6-Al-4V alloy due to lower cutting temperature and BUEs during cryogenic machining conditions (Sun et al. 2015a). Cryogenic machining of titanium alloys has also led to superior surface finish (18%–20%) and higher microhardness (Shokrani, Dhokia, and Newman 2016). Machining of magnesium is always associated with dangerous ignition risk, especially at higher cutting speeds; cryogenic machining of magnesium alloy has been reported to increase the cutting temperature to about 40% to 60% as compared to dry machining process, which increases the sustainability of the process (Dilip Jerold and Pradeep Kumar 2011; Danish et al. 2017; Dinesh, Senthilkumar, and Asokan 2017). Further, beneficial effects in terms of enhanced surface finish and high microhardness were also reported during cryogenic machining of magnesium alloy (Pu et al. 2012; Dinesh, Senthilkumar, and Asokan 2017). Similar beneficial effects of cryogenic machining were also reported for machining of aluminum alloys (Le Coz et al. 2012; Kaynak, Lu, and Jawahir 2014; Zhang et al. 2015; Jawahir et al. 2016). In addition to these positive effects, the liquefied gas (normally liquid nitrogen) that is used for cryogenic machining process spontaneously evaporates into the atmosphere during the machining operation, which makes the process clean and reduces the cost for waste management (Jawahir et al. 2016).

1.7   Conclusions

Manufacturing industries have been increasingly attracted toward using light alloys, especially aluminum, magnesium, and titanium alloys, due to their special property of high strength-to-weight ratio. Light alloys have been considered as a promising candidate for a variety of applications, such as in automobiles, aeronautical, electronics, and also in biomedical applications. In this chapter, the machinability of light alloys was comprehensively discussed, especially the cutting temperature, cutting forces, tools, and tool wear.

One of the main problems that normally occurred during the machining of light alloys is the high cutting temperature. In the case of aluminum, it can increase the adhesion of work material on cutting tool edge, which can result in high cutting forces and poor surface finish. While machining magnesium alloys, high temperature can also lead to ignition risks, particularly at high cutting speeds. For titanium, in addition to the adhesion of work material on the tool face, the cutting temperature can be very high, which can initiate a chemical reaction between the tool and work material. Proper machining parameter, along with machining conditions, can decrease the cutting temperature during the machining operation. A low cutting speed and high feed rate are commonly recommended. Furthermore, using liquefied gases can also diminish this problem significantly.

Cutting forces are generally low while machining light alloys, especially for aluminum and magnesium alloys. However, when machining light alloys having hard particles such as silicon or ceramic particles, the cutting forces can suddenly increase due to the contact of the tool with these hard particles. These hard particles can also give rise to the cutting temperature, which can lead to high BUEs, which further results in an increase in tool wear. Therefore, it is recommended that the tool should have high thermal conductivity and high dimensional stability, especially during machining of titanium alloys. During machining of titanium alloy, the tool material and coating should be chosen such that it should not react with the titanium alloys at high temperature (around 500°C). A tool with high hardness like the diamond cutting tool or PCD could be a good option for machining light alloys at high cutting speeds. Proper selection of machining parameters, tool geometry, and tool material can enhance the machinability of the alloy. In addition to this, employing cryogenic condition can also significantly increase the efficiency of the process of machining light alloys. It not only reduces the cutting temperature but also increases the tool life, surface quality, and microhardness and reduces the cutting forces.

References

Abhang, L. B., and M. Hameedullah. 2010. "Chip–Tool Interface Temperature Prediction Model for Turning Process." *International Journal of Engineering Science and Technology* 2 (4): 382–93.

Abukhshim, N. A., P. T. Mativenga, and M. A. Sheikh. 2006. "Heat Generation and Temperature Prediction in Metal Cutting: A Review and Implications for High Speed Machining." *International Journal of Machine Tools and Manufacture* 46 (7–8): 782–800. doi:10.1016/j.ijmachtools.2005.07.024.

Alniak, M. O., and F. Bedir. 2003. "Changes of Grain Sizes and Flow Stresses of AA2014 and AA6063 Aluminum Alloys at High Temperatures in Various Strain Rates." *Turkish Journal of Engineering and Environmental Sciences* 27 (1): 59–64.

Andrewes, C. J. E., H. Y. Feng, and W. M. Lau. 2000. "Machining of an Aluminum/SiC Composite Using Diamond Inserts." *Journal of Materials Processing Technology* 102 (1): 25–29. doi:10.1016/S0924-0136(00)00425-8.

Anilchandra, A. R., and M. K. Surappa. 2010. "Influence of Tool Rake Angle on the Quality of Pure Magnesium Chip-Consolidated Product." *Journal of Materials Processing Technology* 210 (3): 423–28. doi:10.1016/j.jmatprotec.2009.10.002.

Antonialli, A. I. S., A. E. Diniz, and R. Pederiva. 2010. "Vibration Analysis of Cutting Force in Titanium Alloy Milling." *International Journal of Machine Tools and Manufacture* 50 (1): 65–74. doi:10.1016/j.ijmachtools.2009.09.006.

Arumugam, P. U., A. P. Malshe, and S. A. Batzer. 2006. "Dry Machining of Aluminum-Silicon Alloy Using Polished CVD Diamond-Coated Cutting Tools Inserts." *Surface and Coatings Technology* 200 (11): 3399–403. doi:10.1016/j.surfcoat.2005.08.127.

ASM. 1989. *ASM Handbook-Vol 16-Machining*. Ohio: ASM International.

Basha, I., E. Ho, C. Yousuff, and N. Hamid. 2017. "Towards Multiplex Molecular Diagnosis—A Review of Microfluidic Genomics Technologies." *Micromachines* 8 (9): 266. doi:10.3390/mi8090266.

Beranoagirre, A., and L. N. López De Lacalle. 2011. "Topography Prediction on Milling of Emerging Aeronautical Ti Alloys." *Physics Procedia* 22: 136–43. doi:10.1016/j.phpro.2011.11.022.

Biermann, D., and M. Heilmann. 2010. "Improvement of Workpiece Quality in Face Milling of Aluminum Alloys." *Journal of Materials Processing Technology* 210 (14): 1968–75. doi:10.1016/j.jmatprotec.2010.07.010.

Bishop, D. P., J. R. Cahoon, M. C. Chaturvedi, G. J. Kipouros, and W. F. Caley. 2000. "On Enhancing the Mechanical Properties of Aluminum P/M Alloys." *Materials Science and Engineering A* 290 (1–2): 16–24. doi:10.1016/S0921 -5093(00)00957-6.

Braga, D. U., A. E. Diniz, G. W. A. Miranda, and N. L. Coppini. 2002. "Using a Minimum Quantity of Lubricant (MQL) and a Diamond Coated Tool in the Drilling of Aluminum-Silicon Alloys." *Journal of Materials Processing Technology* 122 (1): 127–38. doi:10.1016/S0924-0136(01)01249-3.

Byrne, G., D. Dornfeld, and B. Denkena. 2003. "Advancing Cutting Technology." *CIRP Annals—Manufacturing Technology* 52 (2): 483–507. doi:10.1016/S0007 -8506(07)60200-5.

Campatelli, G., and A. Scippa. 2012. "Prediction of Milling Cutting Force Coefficients for Aluminum 6082-T4." *Procedia CIRP* 1: 563–68. doi:10.1016/j .procir.2012.04.100.

Campbell, F. 2006. *Manufacturing Technology for Aerospace Structural Materials. Manufacturing Technology for Aerospace Structural Materials*. New York: Elsevier. doi:10.1016/B978-1-85617-495-4.X5000-8.

Carou, D., E. M. Rubio, and J. P. Davim. 2015. "Machinability of Magnesium and Its Alloys: A Review." *Traditional Machining Processes* 133–52. doi:10.1007/978 -3-662-45088-8_5.

Carou, D., E. M. Rubio, B. Agustina, and M. M. Marín. 2017. "Experimental Study for the Effective and Sustainable Repair and Maintenance of Bars Made of Ti-6Al-4V Alloy. Application to the Aeronautic Industry." *Journal of Cleaner Production* 164 (October): 465–75. doi:10.1016/j.jclepro.2017.06.095.

Castro, G., F. A. Almeida, F. J. Oliveira, A. J. S. Fernandes, J. Sacramento, and R. F. Silva. 2008. "Dry Machining of Silicon-Aluminium Alloys with CVD Diamond Brazed and Directly Coated Si_3N_4 Ceramic Tools." *Vacuum* 82 (12): 1407–10. doi:10.1016/j.vacuum.2008.03.042.

Chambers, A. R. 1996. "The Machinability of Light Alloy MMCs." *Composites Part A: Applied Science and Manufacturing* 27 (2): 143–47. doi:10.1016/1359 -835X(95)00001-I.

Charles, J. A., F. A. A. Crane, and J. A. G. Furness. 1997. *Selection and Use of Engineering Materials. Selection and Use of Engineering Materials*. Oxford, UK: Butterworth-Heinemann. doi:10.1016/B978-075063277-5/50023-4.

Che-Haron, C. H., and A. Jawaid. 2005. "The Effect of Machining on Surface Integrity of Titanium Alloy Ti-6% Al-4% v." *Journal of Materials Processing Technology* 166 (2): 188–92. doi:10.1016/j.jmatprotec.2004.08.012.

Ciftci, I., M. Turker, and U. Seker. 2004. "Evaluation of Tool Wear When Machining SiCp-Reinforced Al-2014 Alloy Matrix Composites." *Materials and Design* 25 (3): 251–55. doi:10.1016/j.matdes.2003.09.019.

Clayton, C. R. 1987. "Materials Science and Engineering: An Introduction." *Materials Science and Engineering* 94: 266–67. doi:10.1016/0025-5416(87)90343-0.

Coelho, R. T., S. Yamada, D. K. Aspinwall, and M. L. H. Wise. 1995. "The Application of Polycrystalline Diamond (PCD) Tool Materials When Drilling and Reaming Aluminium Based Alloys Including MMC." *International Journal of Machine Tools and Manufacture* 35 (5): 761–74. doi:10.1016/0890-6955(95)93044-7.

Danish, M., T. L. Ginta, A. U. Alkali, and M. Yasir. 2016. "Thermal Analysis of Cryogenic Machining of Mg Alloy Using FEM." *ARPN Journal of Engineering and Applied Sciences* 11 (8): 5134–38.

Danish, M., T. L. Ginta, K. Habib, A. M. A. Rani, and B. B. Saha. 2018. "Effect of Cryogenic Cooling on the Heat Transfer during Turning of AZ31C Magnesium Alloy." *Heat Transfer Engineering*. doi: 10.1080/01457632.2018.1450345.

Danish, M., T. Lenggo Ginta, K. Habib, D. Carou, A. M. A. Rani, and B. Baran Saha. 2017. "Thermal Analysis during Turning of AZ31 Magnesium Alloy under Dry and Cryogenic Conditions." *The International Journal of Advanced Manufacturing Technology* 91 (5–8): 2855–68. doi:10.1007/s00170-016-9893-5.

Danish, M., T. L. Ginta, and B. Ari Wahjoedi. 2016. "Enhanced Functional Properties of Mg Alloys by Cryogenic Machining." *International Journal of Applied Engineering Research* 11 (7): 5055–59.

Dasch, J. M., C. C. Ang, C. A. Wong, Y. T. Cheng, A. M. Weiner, L. C. Lev, and E. Konca. 2006. "A Comparison of Five Categories of Carbon-Based Tool Coatings for Dry Drilling of Aluminum." *Surface and Coatings Technology* 200 (9): 2970–77. doi:10.1016/j.surfcoat.2005.04.025.

Davies, G. 2003. "Future Trends in Automotive Body Materials." *Materials for Automobile Bodies* 252–69. doi:10.1016/B978-0-08-096979-4.00009-8.

Davis, J. R. 2007. "Microstructures of Aluminum Alloys." In Dawis J. R. (ed), *ASM Specialty Handbook Aluminum and Aluminum Alloys*. Ohio: ASM International 10. https://www.asminternational.org.

Demir, H., and S. Gündüz. 2009. "The Effects of Aging on Machinability of 6061 Aluminium Alloy." *Materials and Design* 30 (5): 1480–83. doi:10.1016/j.matdes.2008.08.007.

Denkena, B., and A. Lucas. 2007. "Biocompatible Magnesium Alloys as Absorbable Implant Materials Adjusted Surface and Subsurface Properties by Machining Processes." *CIRP Annals—Manufacturing Technology* 56 (1): 113–16. doi:10.1016/j.cirp.2007.05.029.

Destefani, J. D. 1990. "Introduction to Titanium and Titanium Alloys." In *ASM Handbook, Volume 2: Properties and Selection: Nonferrous Alloys and Special-Purpose Materials*, 586–91.

Dilip Jerold, B., and M. Pradeep Kumar. 2011. "Experimental Investigation of Turning AISI 1045 Steel Using Cryogenic Carbon Dioxide as the Cutting Fluid." *Journal of Manufacturing Processes* 13 (2): 113–19. doi:10.1016/j.jmapro.2011.02.001.

Dilip Jerold, B., and M. Pradeep Kumar. 2012. "Experimental Comparison of Carbon-Dioxide and Liquid Nitrogen Cryogenic Coolants in Turning of AISI 1045 Steel." *Cryogenics* 52 (10): 569–74. doi:10.1016/j.cryogenics.2012.07.009.

Dinesh, S., V. Senthilkumar, and P. Asokan. 2017. "Experimental Studies on the Cryogenic Machining of Biodegradable ZK60 Mg Alloy Using Micro-Textured Tools." *Materials and Manufacturing Processes* 32 (9): 979–87. doi:10.1080/10426914.2016.1221096.

Dinesh, S., V. Senthilkumar, P. Asokan, and D. Arulkirubakaran. 2015. "Effect of Cryogenic Cooling on Machinability and Surface Quality of Bio-Degradable ZK60 Mg Alloy." *Materials and Design* 87: 1030–36. doi:10.1016/j.matdes.2015.08.099.

Dornfeld, D., and S. Min. 2010. "A Review of Burr Formation in Machining." *Burrs—Analysis, Control and Removal.* 3–11 doi:10.1007/978-3-642-00568-8.

Dwivedi, D. K., A. Sharma, and T. V. Rajan. 2008. "Machining of LM13 and LM28 Cast Aluminium Alloys: Part I." *Journal of Materials Processing Technology* 196 (1–3): 197–204. doi:10.1016/j.jmatprotec.2007.05.032.

Elgallad, E. M., F. H. Samuel, A. M. Samuel, and H. W. Doty. 2010. "Machinability Aspects of New Al-Cu Alloys Intended for Automotive Castings." *Journal of Materials Processing Technology* 210 (13): 1754–66. doi:10.1016/j.jmatprotec.2010.06.006.

Ezugwu, E. O. 2005. "Key Improvements in the Machining of Difficult-to-Cut Aerospace Superalloys." *International Journal of Machine Tools and Manufacture* 45 (12–13): 1353–67. doi:10.1016/j.ijmachtools.2005.02.003.

Ezugwu, E. O., and Z. M. Wang. 1997. "Titanium Alloys and Their Machinability—A Review." *Journal of Materials Processing Technology* 68 (3): 262–74. doi:10.1016/S0924-0136(96)00030-1.

Ezugwu, E. O., R. B. Da Silva, J. Bonney, and Á. R. MacHado. 2005. "Evaluation of the Performance of CBN Tools When Turning Ti-6Al-4V Alloy with High Pressure Coolant Supplies." *International Journal of Machine Tools and Manufacture* 45 (9): 1009–14. doi:10.1016/j.ijmachtools.2004.11.027.

Fang, N., and Q. Wu. 2005. "The Effects of Chamfered and Honed Tool Edge Geometry in Machining of Three Aluminum Alloys." *International Journal of Machine Tools and Manufacture* 45 (10): 1178–87. doi:10.1016/j.ijmachtools.2004.12.003.

Fraker, A. C., A. W. Ruff, P. Sung, A. C. Van Orden, and K. M. Speck. 1983. "Surface Preparation and Corrosion Behavior of Titanium Alloys for Surgical Implants." In *Titanium Alloys in Surgical Implants.* West Conshohocken, PA: ASTM International, pp. 206–14. doi:10.1520/STP28944S.

Froes, F. H., D. Eliezer, and E. Aghion. 1998. "The Science, Technology, and Applications of Magnesium." *JOM Journal of the Minerals, Metals and Materials Society.* doi:10.1007/s11837-998-0411-6.

Fuh, K.-H., and H.-Y. Chang. 1997. "An Accuracy Model for the Peripheral Milling of Aluminum Alloys Using Response Surface Design." *Journal of Materials Processing Technology* 72 (1): 42–47. doi:10.1016/S0924-0136(97)00127-1.

Gariboldi, E. 2003. "Drilling a Magnesium Alloy Using PVD Coated Twist Drills." *Journal of Materials Processing Technology* 134 (3): 287–95. doi:10.1016/S0924-0136(02)01111-1.

Gómez-Parra, A., M. Álvarez-Alcón, J. Salguero, M. Batista, and M. Marcos. 2013. "Analysis of the Evolution of the Built-Up Edge and Built-Up Layer Formation Mechanisms in the Dry Turning of Aeronautical Aluminium Alloys." *Wear* 302 (1–2): 1209–18. doi:10.1016/j.wear.2012.12.001.

Gray, J. E. E., and B. Luan. 2002. "Protective Coatings on Magnesium and Its Alloys—A Critical Review." *Journal of Alloys and Compounds* 336: 88–113. doi:10.1016/S0925-8388(01)01899-0.

Grzesik, W. 2008. "Machinability of Engineering Materials." In *Advanced Machining Processes of Metallic Materials* 241–64. doi:10.1016/B978-0-444-63711-6.00013-2.

Guimu, Z., Y. Chao, S. R. Chen, and A. Libao. 2003. "Experimental Study on the Milling of Thin Parts of Titanium Alloy (TC4)." In *Journal of Materials Processing Technology* 138:489–93. doi:10.1016/S0924-0136(03)00126-2.

Hamade, R. F., and F. Ismail. 2005. "A Case for Aggressive Drilling of Aluminum." *Journal of Materials Processing Technology* 166 (1): 86–97. doi:10.1016/j.jmatprotec.2004.07.099.

Haron, C. H. Che, A. Ginting, and H. Arshad. 2007. "Performance of Alloyed Uncoated and CVD-Coated Carbide Tools in Dry Milling of Titanium Alloy Ti-6242S." *Journal of Materials Processing Technology* 185 (1–3): 77–82. doi:10.1016/j.jmatprotec.2006.03.135.

Hatch J. E. Aluminum Association, American Society for Metals. 1984. "Aluminum Properties and Physical Metallurgy." Ohio: Aluminum Association Inc. and ASM International, 424 p. doi:10.1361/appm1984p001.

Hong, S. Y., and Y. Ding. 2001. "Cooling Approaches and Cutting Temperatures in Cryogenic Machining of Ti-6Al-4V." *International Journal of Machine Tools and Manufacture* 41 (10): 1417–37. doi:10.1016/S0890-6955(01)00026-8.

Hong, S. Y., Y. Ding, and W.-C. Jeong. 2001. "Friction and Cutting Forces in Cryogenic Machining of Ti-6Al-4V." *International Journal of Machine Tools and Manufacture* 41 (15): 2271–85. doi:10.1016/S0890-6955(01)00029-3.

Huda, Z., N. I. Taib, and T. Zaharinie. 2009. "Characterization of 2024-T3: An Aerospace Aluminum Alloy." *Materials Chemistry and Physics* 113 (2–3): 515–17. doi:10.1016/j.matchemphys.2008.09.050.

Jawahir, I. S., H. Attia, D. Biermann, J. Duflou, F. Klocke, D. Meyer, S. T. Newman et al. 2016. "CIRP Annals—Manufacturing Technology Cryogenic Manufacturing Processes." *CIRP Annals—Manufacturing Technology* 65: 713–36. doi:10.1016/j.cirp.2016.06.007.

Jawaid, A., C. H. Che-Haron, and A. Abdullah. 1999. "Tool Wear Characteristics in Turning of Titanium Alloy Ti-6246." *Journal of Materials Processing Technology* 92–93: 329–34. doi:10.1016/S0924-0136(99)00246-0.

Jawaid, A., S. Sharif, and S. Koksal. 2000. "Evaluation of Wear Mechanisms of Coated Carbide Tools When Face Milling Titanium Alloy." *Journal of Materials Processing Technology* 99 (1): 266–74. doi:10.1016/S0924-0136(99)00438-0.

Kannan, S., and H. A. Kishawy. 2008. "Tribological Aspects of Machining Aluminium Metal Matrix Composites." *Journal of Materials Processing Technology* 198 (1–3): 399–406. doi:10.1016/j.jmatprotec.2007.07.021.

Kaynak, Y., T. Lu, and I. S. Jawahir. 2014. "Cryogenic Machining-Induced Surface Integrity: A Review and Comparison with Dry, MQL, and Flood-Cooled Machining." *Machining Science and Technology* 18 (2): 149–98. doi:10.1080/10910344.2014.897836.

Kelly, J. F., and M. G. Cotterell. 2002. "Minimal Lubrication Machining of Aluminium Alloys." *Journal of Materials Processing Technology* 120 (1–3): 327–34. doi:10.1016/S0924-0136(01)01126-8.

Kilic, D. S., and S. Raman. 2007. "Observations of the Tool–Chip Boundary Conditions in Turning of Aluminum Alloys." *Wear* 262 (7–8): 889–904. doi:10.1016/j.wear.2006.08.019.

Kipouros, G. J., and D. R. Sadoway. 2001. "A Thermochemical Analysis of the Production of Anhydrous $MgCl_2$." *Journal of Light Metals* 1 (2): 111–17. doi:10.1016/S1471-5317(01)00004-9.

Kishawy, H. A., M. Dumitrescu, E. G. Ng, and M. A. Elbestawi. 2005. "Effect of Coolant Strategy on Tool Performance, Chip Morphology and Surface Quality during High-Speed Machining of A356 Aluminum Alloy." *International Journal of Machine Tools and Manufacture* 45 (2): 219–27. doi:10.1016/j.ijmachtools.2004.07.003.

Kitagawa, T., A. Kubo, and K. Maekawa. 1997. "Temperature and Wear of Cutting Tools in High-Speed Machining of Inconel 718 and $Ti_6Al_6V_2Sn$." *Wear* 202 (2): 142–48. doi:10.1016/S0043-1648(96)07255-9.

Kubiak, M., W. Piekarska, S. Stano, J. C. Ion, S. Stano, L. Quintino, R. Miranda, U. Dilthey, D. Iordachescu, and M. Banasik. 2012. "Laser Processing of Engineering Materials." *Advanced Structured Materials* 83: 33–57. doi:10.1016/j.ijheatmasstransfer.2014.12.052.

Kulekci, M. K. 2008. "Magnesium and Its Alloys Applications in Automotive Industry." *The International Journal of Advanced Manufacturing Technology* 39 (9–10): 851–65. doi:10.1007/s00170-007-1279-2.

Lane, B. M., M. Shi, T. A. Dow, and R. Scattergood. 2010. "Diamond Tool Wear When Machining Al6061 and 1215 Steel." *Wear* 268 (11–12): 1434–41. doi:10.1016/j.wear.2010.02.019.

Larbi, S., S. Djebali, and A. Bilek. 2015. "Study of High Speed Machining by Using Split Hopkinson Pressure Bar." *Procedia Engineering* 114: 314–21. doi:10.1016/j.proeng.2015.08.074.

Le Coz, G., M. Marinescu, A. Devillez, D. Dudzinski, and L. Velnom. 2012. "Measuring Temperature of Rotating Cutting Tools: Application to MQL Drilling and Dry Milling of Aerospace Alloys." *Applied Thermal Engineering* 36 (1): 434–41. doi:10.1016/j.applthermaleng.2011.10.060.

Lei, S., and W. Liu. 2002. "High-Speed Machining of Titanium Alloys Using the Driven Rotary Tool." *International Journal of Machine Tools and Manufacture* 42 (6): 653–61. doi:10.1016/S0890-6955(02)00012-3.

Leyens, C., and M. Peters. 2003. *Titanium and Titanium Alloys. Titanium and Titanium Alloys.* Wienheim, Germany: Wiley-VCH. doi:10.1002/3527602119.

Liang, Q., Y. K. Vohra, and R. Thompson. 2008. "High Speed Continuous and Interrupted Dry Turning of A390 Aluminum/Silicon Alloy Using Nanostructured Diamond Coated WC-6wt.% Cobalt Tool Inserts by MPCVD." *Diamond and Related Materials* 17 (12): 2041–47. doi:10.1016/j.diamond.2008.06.008.

Liew, W. Y. H., I. M. Hutchings, and J. A. Williams. 1998. "Friction and Lubrication Effects in the Machining of Aluminium Alloys." *Tribology Letters* 5 (1998): 117–22.

Liu, J., M. Chung, and S. Park. 2014. "Calculation of Two-Phase Convective Heat Transfer Coefficients of Cryogenic Nitrogen Flow in Plate-Fin Type Heat

Exchangers." *Heat Transfer Engineering* 35 (6–8): 674–84. doi:10.1080/01457
632.2013.837376.

López De Lacalle, L. N., J. Pérez, J. I. Llorente, and J. A. Sánchez. 2000. "Advanced
Cutting Conditions for the Milling of Aeronautical Alloys." *Journal of Materials Processing Technology* 100 (1): 1–11. doi:10.1016/S0924-0136(99)00372-6.

Machado, A. R., and J. Wallbank. 1997. "The Effect of Extremely Low Lubricant Volumes in Machining." *Wear* 210 (1–2): 76–82. doi:10.1016/S0043
-1648(97)00059-8.

Mackerle, J. 1999. "Finite-Element Analysis and Simulation of Machining: A Bib-
liography (1976–1996)." *Journal of Materials Processing Technology* 86 (1–3):
17–44. doi:10.1016/S0924-0136(98)00227-1.

Manna, A., and B. Bhattacharyya. 2002. "A Study on Different Tooling Systems
during Machining of Al/SiC-MMC." *Journal of Materials Processing Technology* 123 (3): 476–82. doi:10.1016/S0924-0136(02)00127-9.

Matweb. 2017. "Overview of Materials for Alpha/beta Titanium Alloy." Accessed
September 10, 2017. http://www.matweb.com/search/DataSheet.aspx?Mat
GUID=4dac23c848db4780a067fd556906cae6&ckck=1.

McDonough, W. F., and S. S. Sun. 1995. "The Composition of the Earth." *Chemical
Geology* 120 (3–4): 223–53. doi:10.1016/0009-2541(94)00140-4.

Menezes, J., M. A. Rubeo, K. Kiran, A. Honeycutt, and T. L. Schmitz. 2016. "Productivity Progression with Tool Wear in Titanium Milling." *Procedia Manufacturing* 5: 427–41. doi:10.1016/j.promfg.2016.08.036.

Miller, W. S., L. Zhuang, J. Bottema, A. J. Wittebrood, P. De Smet, A. Haszler, and
Ai Vieregge. 2000. "Recent Development in Aluminium Alloys for the
Automotive Industry." *Materials Science and Engineering: A* 280 (1): 37–49.
doi:10.1016/S0921-5093(99)00653-X.

Mitsuishi, M., J. Cao, P. Bártolo, D. Friedrich, A. J. Shih, K. Rajurkar, N. Sugita,
and K. Harada. 2013. "Biomanufacturing." *CIRP Annals* 62 (2): 585–606.
doi:10.1016/j.cirp.2013.05.001.

Mordike, B. L., and T. Ebert. 2001. "Magnesium Properties—Applications—
Potential." *Materials Science and Engineering A* 302 (1): 37–45. doi:10.1016/S0921
-5093(00)01351-4.

Ng, E.-G., D. K. Aspinwall, D. Brazil, and J. Monaghan. 1999. "Modelling of
Temperature and Forces When Orthogonally Machining Hardened
Steel." *International Journal of Machine Tools and Manufacture* 39: 885–903.
doi:10.1016/S0890-6955(98)00077-7.

Ng, E.-G., D. Szablewski, M. Dumitrescu, M. A. Elbestawi, and J. H. Sokolowski.
2004. "High Speed Face Milling of a Aluminium Silicon Alloy Casting."
CIRP Annals—Manufacturing Technology 53 (1): 69–72. doi:10.1016/S0007
-8506(07)60647-7.

Ng, C. K., S. N. Melkote, M. Rahman, and A. Senthil Kumar. 2006. "Experimental
Study of Micro- and Nano-Scale Cutting of Aluminum 7075-T6." *International Journal of Machine Tools and Manufacture* 46 (9): 929–36. doi:10.1016/j
.ijmachtools.2005.08.004.

Nouari, M., G. List, F. Girot, and D. Coupard. 2003. "Experimental Analysis and
Optimisation of Tool Wear in Dry Machining of Aluminium Alloys." *Wear*
255 (7–12): 1359–68. doi:10.1016/S0043-1648(03)00105-4.

Nourbakhsh, F., K. P. Rajurkar, A. P. Malshe, and J. Cao. 2013. "Wire Electro-Discharge Machining of Titanium Alloy." *Procedia CIRP* 5: 13–18. doi:10.1016/j.procir.2013.01.003.

Oliaei, S. N. B., and Y. Karpat. 2016. "Investigating the Influence of Built-up Edge on Forces and Surface Roughness in Micro Scale Orthogonal Machining of Titanium Alloy Ti$_6$Al$_4$V." *Journal of Materials Processing Technology* 235 (September): 28–40. doi:10.1016/j.jmatprotec.2016.04.010.

Outeiro, J. C., F. Rossi, G. Fromentin, G. Poulachon, G. Germain, and A. C. Batista. 2013. "Process Mechanics and Surface Integrity Induces by Dry and Cryogenic Machining of AZ31B-0 Magnesium Alloy." *Procedia CIRP* 8: 487–92.

Ozcatalbas, Y. 2003. "Chip and Built-up Edge Formation in the Machining of in Situ Al$_4$C$_3$-Al Composite." *Materials and Design* 24 (3): 215–21. doi:10.1016/S0261-3069(02)00146-2.

Pascual Serrano, D., C. Vera Pasamontes, and R. Girón Moreno. 2016. "Animal Models of Pain. A Critical Review." *DOLOR* 31 (2): 70–76. doi:10.1017/CBO9781107415324.004.

Paulo Davim, J., and A. Monteiro Baptista. 2000. "Relationship between Cutting Force and PCD Cutting Tool Wear in Machining Silicon Carbide Reinforced Aluminum." *Journal of Materials Processing Technology* 103 (3): 417–23. doi:10.1016/S0924-0136(00)00495-7.

Polini, R., F. Casadei, P. D'Antonio, and E. Traversa. 2003. "Dry Turning of Alumina/aluminum Composites with CVD Diamond Coated Co-Cemented Tungsten Carbide Tools." *Surface and Coatings Technology* 166 (2–3): 127–34. doi:10.1016/S0257-8972(02)00775-2.

Polmear, I. J. 2005. *Light Alloys : From Traditional Alloys to Nanocrystals*. New York: Elsevier.

Pu, Z., J. C. Outeiro, A. C. Batista, O. W. Dillon, D. A. Puleo, and I. S. Jawahir. 2012. "Enhanced Surface Integrity of AZ31B Mg Alloy by Cryogenic Machining towards Improved Functional Performance of Machined Components." *International Journal of Machine Tools and Manufacture* 56 (May): 17–27. doi:10.1016/j.ijmachtools.2011.12.006.

Rawangwong, S., J. Chatthong, W. Boonchouytan, and R. Burapa. 2013. "An Investigation of Optimum Cutting Conditions in Face Milling Aluminum Semi Solid 2024 Using Carbide Tool." *Energy Procedia* 34: 854–62. doi:10.1016/j.egypro.2013.06.822.

Rubio, E. M., J. L. Valencia, D. Carou, and A. J. Saa. 2012. "Inserts Selection for Intermittent Turning of Magnesium Pieces." *Applied Mechanics and Materials* 1581: 217–19. doi:http://dx.doi.org/10.4028/www.scientific.net/AMM.217-219.1581.

Saglam, H., F. Unsacar, and S. Yaldiz. 2006. "Investigation of the Effect of Rake Angle and Approaching Angle on Main Cutting Force and Tool Tip Temperature." *International Journal of Machine Tools and Manufacture* 46 (2): 132–41. doi:10.1016/j.ijmachtools.2005.05.002.

Salguero, J., M. Batista, M. Calamaz, F. Girot, and M. Marcos. 2013. "Cutting Forces Parametric Model for the Dry High Speed Contour Milling of Aerospace Aluminium Alloys." *Procedia Engineering* 63:735–42. doi:10.1016/j.proeng.2013.08.215.

Santos, M. C., A. R. Machado, W. F. Sales, M. A. S. Barrozo, and E. O. Ezugwu. 2016. "Machining of Aluminum Alloys: A Review." *The International Journal of Advanced Manufacturing Technology* 86 (9–12): 3067–80. doi:10.1007/s00170-016-8431-9.

Scintilla, L. D., and L. Tricarico. 2013. "Experimental Investigation on Fiber and CO_2 Inert Gas Fusion Cutting of AZ31 Magnesium Alloy Sheets." *Optics and Laser Technology* 46 (1): 42–52. doi:10.1016/j.optlastec.2012.04.026.

Shaw, M. C. 2005. *Metal Cutting Principles: Chapter 3. Oxford Series on Advanced Manufacturing*. United Kingdom: Oxford University Press.

Shen, N., H. Ding, Z. Pu, I. S. Jawahir, and T. Jia. 2017. "Enhanced Surface Integrity from Cryogenic Machining of AZ31B Mg Alloy: A Physics-Based Analysis with Microstructure Prediction." *Journal of Manufacturing Science and Engineering* 139 (6): 61012. doi:10.1115/1.4034279.

Shokrani, A., V. Dhokia, and S. T. Newman. 2012. "Environmentally Conscious Machining of Difficult-to-Machine Materials with Regard to Cutting Fluids." *International Journal of Machine Tools and Manufacture* 57: 83–101. doi:10.1016/j.ijmachtools.2012.02.002.

Shokrani, A., V. Dhokia, and S. T. Newman. 2016. "Investigation of the Effects of Cryogenic Machining on Surface Integrity in CNC End Milling of Ti-6Al-4V Titanium Alloy." *Journal of Manufacturing Processes* 21: 172–79. doi:10.1016/j.jmapro.2015.12.002.

Sridhar, B. R., G. Devananda, K. Ramachandra, and R. Bhat. 2003. "Effect of Machining Parameters and Heat Treatment on the Residual Stress Distribution in Titanium Alloy IMI-834." *Journal of Materials Processing Technology* 139 (1–3 SPEC): 628–34. doi:10.1016/S0924-0136(03)00612-5.

Su, Y., N. He, L. Li, and X. L. Li. 2006. "An Experimental Investigation of Effects of Cooling/lubrication Conditions on Tool Wear in High-Speed End Milling of Ti-6Al-4V." *Wear* 261 (7–8): 760–66. doi:10.1016/j.wear.2006.01.013.

Sun, S., M. Brandt, S. Palanisamy, and M. S. Dargusch. 2015a. "Effect of Cryogenic Compressed Air on the Evolution of Cutting Force and Tool Wear during Machining of Ti-6Al-4V Alloy." *Journal of Materials Processing Technology* 221 (July): 243–54. doi:10.1016/j.jmatprotec.2015.02.017.

Sun, Y., B. Huang, D. A. Puleo, and I. S. Jawahir. 2015b. "Enhanced Machinability of Ti-5553 Alloy from Cryogenic Machining: Comparison with MQL and Flood-Cooled Machining and Modeling." *Procedia CIRP* 31: 477–82. doi:10.1016/j.procir.2015.03.099.

Tan, E., and B. Ogel. 2007. "Influence of Heat Treatment on the Mechanical Properties of AA6066 Alloy." *Turkish Journal of Engineering and Environmental Sciences* 31: 53–60.

Tang, Z. T., Z. Q. Liu, Y. Z. Pan, Y. Wan, and X. Ai. 2009. "The Influence of Tool Flank Wear on Residual Stresses Induced by Milling Aluminum Alloy." *Journal of Materials Processing Technology* 209 (9): 4502–508. doi:10.1016/j.jmatprotec.2008.10.034.

Tharumarajah, A., and P. Koltun. 2007. "Is There an Environmental Advantage of Using Magnesium Components for Light-Weighting Cars?" *Journal of Cleaner Production*. doi:10.1016/j.jclepro.2006.05.022.

Tomac, N., K. Tonnessen, and F. O. Rasch. 1991. "Formation of Flank Build-up in Cutting Magnesium Alloys." *CIRP Annals—Manufacturing Technology* 40 (1): 79–82. doi:10.1016/S0007-8506(07)61938-6.

Toropov, A., S. L. Ko, and B. K. Kim. 2005. "Experimental Study of Burrs Formed in Feed Direction When Turning Aluminum Alloy Al6061-T6." *International Journal of Machine Tools and Manufacture* 45 (9): 1015–22. doi:10.1016/j.ijmachtools.2004.11.031.

Trent, E. M. 1977. *Metal Cutting*. United Kingdom: Butterworth and Co. (Publisher) Ltd.

Troeger, L. P., and E. A. Starke. 2000. "Microstructural and Mechanical Characterization of a Superplastic 6xxx Aluminum Alloy." *Materials Science and Engineering: A* 277 (1–2): 102–13. doi:10.1016/S0921-5093(99)00543-2.

Veiga, C., J. P. Davim, and A. J. R. Loureiro. 2013. "Review on Machinability of Titanium Alloys: The Process Perspective." *Reviews on Advanced Materials Science* 32(2): 148–64.

Venugopal, K. A., S. Paul, and A. B. Chattopadhyay. 2007. "Tool Wear in Cryogenic Turning of Ti-6Al-4V Alloy." *Cryogenics* 47 (1): 12–18. doi:10.1016/j.cryogenics.2006.08.011.

Vijaya R., B. C. Elanchezhian, M. Jaivignesh, S. Rajesh, C. Parswajinan, and A. Siddique Ahmed Ghias. 2014. "Evaluation of Mechanical Properties of Aluminium Alloy-Alumina-Boron Carbide Metal Matrix Composites." *Materials & Design*. 58: 332–38. doi:http://dx.doi.org/10.1016/j.matdes.2014.01.068.

Villeta, M., E. M. Rubio, J. M. Sáenz De Pipaón, and M. A. Sebastián. 2011. "Surface Finish Optimization of Magnesium Pieces Obtained by Dry Turning Based on Taguchi Techniques and Statistical Tests." *Materials and Manufacturing Processes* 26 (12): 1503–10. doi:10.1080/10426914.2010.544822.

Wang, J., Y. B. Liu, J. An, and L. M. Wang. 2008. "Wear Mechanism Map of Uncoated HSS Tools during Drilling Die-Cast Magnesium Alloy." *Wear* 265 (5): 685–91. doi:10.1016/j.wear.2007.12.009.

Wang, Z. G., Y. S. Wong, and M. Rahman. 2005. "High-Speed Milling of Titanium Alloys Using Binderless CBN Tools." *International Journal of Machine Tools and Manufacture* 45 (1): 105–14. doi:10.1016/j.ijmachtools.2004.06.021.

Wang, Z. Y., and K. P. Rajurkar. 2000. "Cryogenic Machining of Hard-to-Cut Materials." *Wear* 239 (2): 168–75. doi:10.1016/S0043-1648(99)00361-0.

Weinert, K., I. Inasaki, J. W. Sutherland, and T. Wakabayashi. 2004. "Dry Machining and Minimum Quantity Lubrication." *CIRP Annals* 53 (2): 511–37. doi:10.1016/S0007-8506(07)60027-4.

Westengen, H., and H. M. M. A. Rashed. 2016. "Magnesium Alloys: Properties and Applications." *Reference Module in Materials Science and Materials Engineering* 302 (1): 37–45. doi:10.1016/B978-0-12-803581-8.02568-6.

Witte, F. 2015. "Reprint of: The History of Biodegradable Magnesium Implants: A Review." *Acta Biomaterialia* 23: S28–40. doi:10.1016/j.actbio.2015.07.017.

Witte, F., F. Feyerabend, P. Maier, J. Fischer, M. Störmer, C. Blawert, W. Dietzel, and N. Hort. 2007. "Biodegradable Magnesium–Hydroxyapatite Metal Matrix Composites." *Biomaterials* 28 (13): 2163–74. doi:10.1016/j.biomaterials.2006.12.027.

Xu, K., B. Zou, Y. Wang, P. Guo, C. Huang, and J. Wang. 2016. "An Experimental Investigation of Micro-Machinability of Aluminum Alloy 2024 Using Ti(C_7N_3)-Based Cermet Micro End-Mill Tools." *Journal of Materials Processing Technology* 235 (September): 13–27. doi:10.1016/j.jmatprotec.2016.04.011.

Yousefi, R., and Y. Ichida. 2000. "A Study on Ultra-High-Speed Cutting of Aluminium Alloy:" *Precision Engineering* 24 (4): 371–76. doi:10.1016/S0141 -6359(00)00048-9.

Yousuff, C., M. Danish, E. Ho, I. K. Basha, and N. Hamid. 2017. "Study on the Optimum Cutting Parameters of an Aluminum Mold for Effective Bonding Strength of a PDMS Microfluidic Device." *Micromachines* 8 (8): 258. doi:10.3390 /mi8080258.

Zaghbani, I., and V. Songmene. 2009. "A Force–Temperature Model Including a Constitutive Law for Dry High Speed Milling of Aluminium Alloys." *Journal of Materials Processing Technology* 209 (5): 2532–44. doi:10.1016/j .jmatprotec.2008.05.050.

Zhang, X., H. Mu, X. Huang, Z. Fu, D. Zhu, and H. Ding. 2015. "Cryogenic Milling of Aluminium-Lithium Alloys: Thermo-Mechanical Modelling towards Fine-Tuning of Part Surface Residual Stress." In *Procedia CIRP* 31: 160–65. doi:10.1016/j.procir.2015.03.055.

Zhao, N., J. Hou, and S. Zhu. 2011. "Chip Ignition in Research on High-Speed Face Milling AM50A Magnesium Alloy." In *2011 2nd International Conference on Mechanic Automation and Control Engineering, MACE 2011—Proceedings*. Inner Mongolia, China, July 15–17, 1102–105, IEEE Xplore. doi:10.1109 /MACE.2011.5987127.

chapter two

Investigation on the built-up edge process when dry machining aeronautical aluminum alloys

*Mohammed Nouari, Badis Haddag,
Abdelhadi Moufki, and Samir Atlati*

Contents

2.1   Introduction

Despite the large number of works on machining aluminum alloys, the understanding of the effect of friction conditions requires further investigations. It is well known that with increasing friction, tool wear increases, especially when dry machining of aluminum alloys. Consequently, the surface integrity and the machined components can be strongly influenced by the tribological conditions. The former can also affect the chip formation process (Mabrouki et al. 2016; Nouari et al. 2005). The accumulation of debris generated by this process and deposited on the tool surface during machining induces the formation the built-up edge (BUE). Several authors describe this phenomenon as a wear mechanism. From an experimental point of view, some authors noted that the BUE is significantly affected by the state of stress around the tool cutting edge and happens under extreme conditions of the contact at the tool–chip interface as high friction, high pressure, and sliding velocity. Kone et al. (2011) reported that this phenomenon can be attributed to the brittle behavior of the machined material and the temperature gradient on the tool surface. Other authors showed that the BUE is the consequence of seizure and sublayer flow material at the tool–chip interface. The strain hardening of the workpiece material promotes the formation of a stagnant build-up around the cutting edge. Iwata and Ueda (1980) analyzed scanning electron microscopy images and observed a local deformation around the BUE area for low-carbon steels. Using electron probe microanalysis and electron diffraction, Ramaswami (1971) and Ohgo (1978) showed that adhesion wear has an effect on the BUE and tool wear, which depends directly on cutting conditions. Also, Selvam and Radhakrishnan (1974) investigated different groove tool profiles and analyzed the effect of generated wear on the BUE adhering to the machined surface.

The evolution of the BUE and built-up layer formation mechanisms during dry turning of aeronautical aluminum alloys was studied by Gomez-Parra et al. (2013) and Sukvitfayawong and Inasaki (1994). These authors proposed a new method for BUE detection by measuring cutting forces. They showed that the tool rake angle increases with BUE formation. Frang et al. (2010) identified a quantitative relationship between the BUE formation and cutting vibrations when machining the 2024-T351 aluminum alloy. To identify domains where BUE appears regularly and irregularly, a statistical parameter was introduced by Senthil Kumar et al. (2006), who observed that the tool material is responsible for the apparition of work material adhesion on the rake face, followed then by the formation of BUE. These authors stated that alumina-based ceramic cutting tools are chemically more stable than high-speed steels and carbides, thus having less tendency to adhere to metals during machining and less tendency to form BUE. Rahim and Sasahara (2011) concluded that lubrication reduces the adhesion of the metal on the cutting tool compared to the dry condition.

More recently, Haddag et al. (2014a, 2014b) showed that tool geometry can have an important effect on tribological behavior, especially on work material adhesion on the rake face. In their analysis, they concluded that the adhesion phenomenon occurs in the concave zones on the grooved rake face and consequently induces the BUE.

The BUE has a significant influence on the wear mechanisms in machining and then on the tool life and surface quality. The main goal of the current work is to investigate the relationship between the BUE formation process and the tribological conditions at the tool–chip and tool–workpiece interfaces when machining ductile materials. The usual aeronautical aluminium alloy AA2024 and the uncoated cemented carbide tool WC-Co were chosen as a tool–workpiece couple. An original method based on time-dependent friction at the tool–work material interface is proposed to predict the BUE formation. A numerical finite element model (based on the Arbitrary Lagrangian Eulerian [ALE] approach) was developed to simulate orthogonal cutting with the commercial Abaqus (2009) code. The contact conditions causing the BUE were highlighted by several simulations. First of all, the proposed model is experimentally validated using machining tests under different cutting conditions. To show the effect of contact conditions on BUE formation, several simulations were performed by varying the local friction gradually (first case) and abruptly (second case). All results obtained from these two studied cases are discussed and fully analyzed to understand the friction change and thermomechanical field evolutions at the cutting zone when the BUE is formed.

2.2   Experimental set-up

To analyze the chip formation process and BUE, different experiments have been done under low and high cutting velocities, from 6 to 500 m/min. The cutting tools are made of cemented tungsten carbide with 6 wt.% of cobalt as bender. Different geometries were considered with rake angles about 0°, 15°, and 30°. For all cutting tools, the flank angle was kept constant at 7°, and the edge radius, at 40 μm. Aluminium alloy AA2024-T351 was chosen as the workpiece material, whose chemical composition is shown in Table 2.1 (List et al. 2005).

2.2.1   Chip morphology and BUE formation

Figure 2.1 shows formed chips obtained with two cutting conditions. For a feed of 0.1 mm, a continuous shape can be clearly observed. For large feeds (about 0.3 mm), a segmented shape is obtained with 0° and 15° tools, but the intensity of segmentation is different. The analysis of these results shows that there is a strong correlation between cutting conditions, cutting parameters (rake angle), and the morphology of the formed chips. The results of Table 2.2 show different conditions that indicate the conditions of the BUE formation (List et al. 2005). The high cutting force for a 0° rake angle and 0.1 mm of undeformed chip thickness explains why the

Table 2.1 Chemical composition (wt.%) of AA2024-T351

Al	Cr	Cu	Fe	Mg	Mn	Si	Ti	Zn
Balanced	Max. 0.1	3.8–4.9	Max. 0.5	1.2–1.8	0.3–0.9	Max. 0.5	Max. 0.15	Max. 0.25

Source: List, G., Nouari, M., Géhin, D., Gomez, S., Manaud, J.P., Le Petitcorps, Y., Girot, F. *Wear*, 259, 1177–1189, 2005.

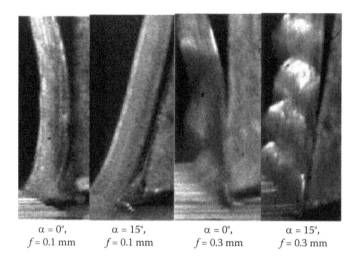

| $\alpha = 0°,$ | $\alpha = 15°,$ | $\alpha = 0°,$ | $\alpha = 15°,$ |
| $f = 0.1$ mm | $f = 0.1$ mm | $f = 0.3$ mm | $f = 0.3$ mm |

Figure 2.1 Continuous and segmented chips obtained with feeds of 0.1 and 0.3 mm and cutting speed of 60 m/min.

Table 2.2 Average cutting and feed forces for a low cutting speed of 60 m/min

Tool-rake angle (°)	Cutting force (N)	Feed force (N)	BUE formation
0	250	150	Yes
15	250	125	No
30	250	100	No
0	500	300	Yes
15	450	250	No
30	400	100	No

Source: List, G., Nouari, M., Géhin, D., Gomez, S., Manaud, J.P., Le Petitcorps, Y., Girot, F. *Wear*, 259, 1177–1189, 2005.

chip meets difficulties in flowing on the tool surface, since the work material flow changes by 90° (from horizontal direction to vertical one on the rake face, as seen in Figure 2.2b). This explains the formation of the BUE in the case of low rake angle.

Figure 2.3 shows the effect of the rake angle and feed on the cutting forces. For low values of feeds (0.05 mm), the rake angle has a neglected effect on the cutting forces, while for a feed of 0.1 mm, the rake angle has an important effect on cutting forces (Figure 2.3b). It can be deduced from Figures 2.2b and 2.3 that the BUE contributes to the increase in cutting forces. This can be assigned to a change in the tribological conditions along the tool rake face.

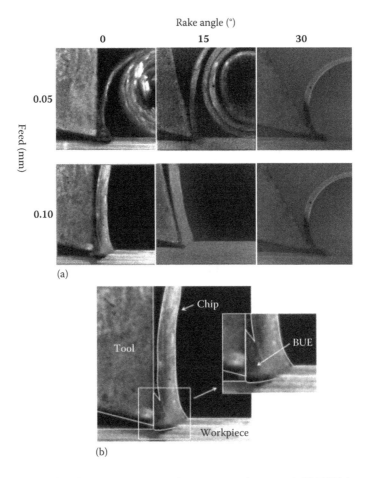

Rake angle (°)

Figure 2.2 (a) Chip formation process for $v_c = 60$ m/min, and (b) BUE formation for $v_c = 60$ m/min, $f = 0.1$ mm, and $\alpha = 0°$.

2.2.2 Evolution of cutting forces and contact length

Experimental results obtained from machining tests showed that cutting and feed forces decrease drastically when the cutting speed increases until the apparition of a stabilized domain at very high cutting velocities (300, 400, and 500 m/min; see Figure 2.4). Two important points can explain the evolution of cutting forces. The first point concerns the combination of the strain rate sensitivity of the work material and the second point is the thermal softening. At low cutting velocities, the temperature rise due to the plastic deformation is not large enough and the work hardening is higher than the softening effect of the work material. In the other side, when the cutting speed increases, a large

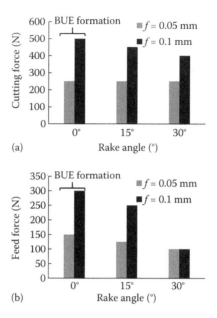

Figure 2.3 Evolution of the cutting forces with the rake angle and feed. (a) Cutting force. (b) Feed force.

part of the plastic work is converted to heat energy, which promotes thermal softening of the work material. The plastic flow stress is then reduced and the cutting forces reduce too. As shown in Figure 2.4, the second mechanism is that the contact length decreases when the cutting speed increases. When the contact length reduces, the friction force at the tool–chip interface reduces too and then causes a decrease in the cutting forces.

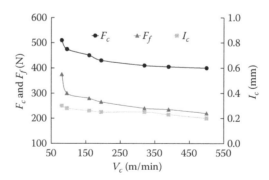

Figure 2.4 Evolution of cutting forces (cutting and feed force) and contact length vs. cutting speed.

2.3   Thermomechanical modelling

2.3.1   Simulation of the chip formation process

By using the Abaqus/Explicit software, a two-dimensional finite element (FE) model has been developed in order to analyze the effects of cutting conditions on the chip formation process and BUE. The cutting tool is supposed to be rigid with thermal behavior. In the workpiece, we define a thin layer of 20 μm, which is deleted as the tool forms the chip. A thermoviscoplastic-damage behavior is considered in the deleted layer and in the uncut chip, with thickness equal to the feed. For the work material below the machined surface, the thermoviscoplastic behavior is given by the Johnson-Cooke law without damage.

The FE results relative to the chip morphology allow examining the pertinence of the work material behavior law. For the cutting conditions in Section 2, the chip morphologies are shown in Figure 2.5. The simulation results are in good agreement with experimental data. Segmented chips are well reproduced in both cases (rake angles 0° and 15°) for the large feed (0.3 mm). As reported in Figure 2.5, chip curvature depends

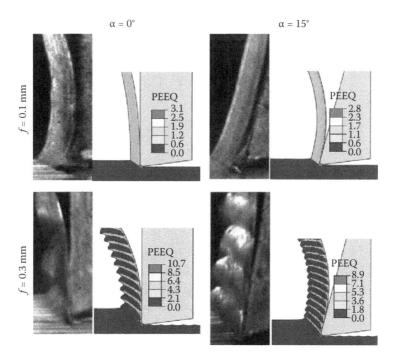

Figure 2.5 Experimental and numerical chip morphologies and equivalent plastic strain (PEEQ) inside the chip segment.

on the tool rake angle. Figure 2.5 also reveals a relative difference with the experimental observations. This difference can be related to the fact that in FE model, the chip curvature is slightly affected by numerical parameters such as FE formulation and mesh density (Haddag et al. 2010).

To study the effect of cutting velocity on the thermomechanical cutting parameters, experimental cutting conditions are simulated by using the FE approach. As reported in Figure 2.6, the predicted cutting forces and tool–chip contact length are globally underestimated for cutting speed from 30 to 500 m/min. One possible explanation of this finding is the impact of the material behavior, represented by the Johnson-Cook model, which may underestimate the work-hardening at moderate and high strain rates.

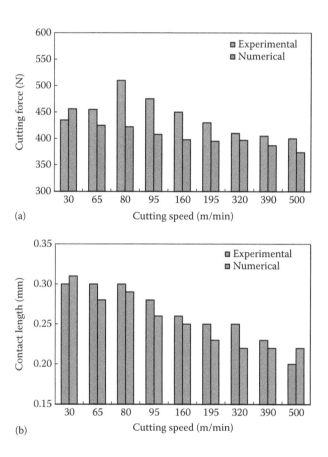

Figure 2.6 (a) Cutting force and (b) tool–chip contact length as a function of cutting speed (feed = 0.1 mm) and rake angle (0°).

2.3.2 Simulation of the BUE process

BUE results from the adhesive layer of work material on the tool rake face and the rounded cutting edge. Thus, its process formation can be assigned to the increase in the friction coefficient μ_{loc} with time. Note that μ_{loc} corresponds to the local friction coefficient (i.e., in the sliding zone), while the friction coefficient deduced from the experimental cutting forces results from the whole tool–chip contact (rake face, cutting edge, and clearance face). For the evolution of μ_{loc}, two cases have been considered: (i) a sudden variation in μ_{loc} and (ii) a progressive change in μ_{loc}. In order to examine the influences of (i) and (ii) on BUE, the following cutting parameters will be analyzed: the plastic strain in the secondary shear zone (SSZ), the material velocity along the tool rake face, and the cutting and feed forces.

To analyze the impact of a sudden change in μ_{loc} on BUE formation, two cases were presented: μ_{loc} is changed from 0.2 to 0.4 and 0.6. The evolution of the tool–work material interaction is presented in Figure 2.7, from which it can be observed that the material velocity field changes with the increase in μ_{loc}. The variation of the material flow is more significant for $\mu_{loc} = 0.6$ and the sticking zone becomes larger. Therefore, the tool–chip interface corresponds to a sticking–sliding contact. For $\mu_{loc} = 0.2$, a negligible sticking area can be observed. Consequently, no BUE is formed and the tool–chip contact is almost completely a sliding contact without any sticking. In addition, the abrupt increase in μ_{loc} induces an augmentation of the SSZ. The SSZ size grows with the BUE. It should be reminded that the sticking contact at the tool rake face results from the thermal softening of the workpiece material. Chip heating is caused by three sources: (i) the viscoplastic deformation in the primary shear zone, (ii) the viscoplastic deformation in the SSZ, and (iii) the frictional heat in the sliding zone of the tool–chip contact. When the temperature is large enough, as

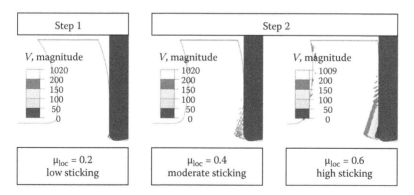

Figure 2.7 Effect of the local friction coefficient μ_{loc} change on the material velocity along the tool rake face.

in the case $\mu_{loc} = 0.6$, the shear stress of the work material at the tool–chip interface becomes lower than the shear stress resulting from the Coulomb friction law ($\tau = \mu_{loc}p$; p is the contact pressure). This results in a sticking contact. From the present analysis, it appears that the tribological conditions in machining, including the formation of BUE and the sticking–sliding contact, depend strongly on the local friction coefficient.

According to Figure 2.8, the sudden increase in the local friction coefficient affects highly the cutting forces during the thermomechanical chip formation process. In fact, when μ_{loc} is increased, the material flow in the chip becomes more difficult, resulting in higher cutting forces. However, it can be observed that the influence of friction is more pronounced for the feed force than for the cutting force, as reported in Figure 2.8. This trend is due to the fact that when the tool rake angle is zero, as in the present work, the tool rake face is parallel to the feed direction. Consequently, the feed force component is directly affected by any modification in the tribological conditions during the chip flow along the tool rake face.

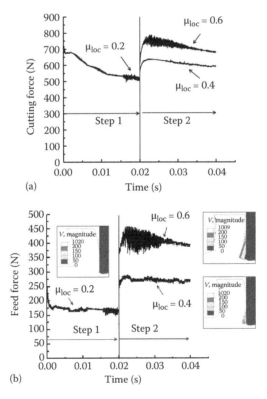

Figure 2.8 Variation of (a) the cutting force and (b) the feed force with time for different values of the local friction coefficient.

The BUE formation can also result from a progressive evolution of the friction coefficient when the chip interacts with the tool. To simulate this case, the local friction coefficient μ_{loc} has been augmented gradually from 0.2 to 1, with an increment of 0.1. As previously, the BUE formation process is analyzed with respect to the following cutting parameters: plastic deformation field, material flow in the chip, and cutting forces.

The tendencies observed in Figure 2.9 confirm the correlation between the evolution of the sticking contact and the variation of the friction conditions at the tool–chip interface. The BUE and the sticking area gradually increase with μ_{loc} because of the thermal softening in the SSZ. The same trend may be deduced from Figure 2.10 relative to the influence of μ_{loc} on the plastic deformation near the cutting edge and the tool rake face.

The increase in BUE with μ_{loc} makes the cutting and feed forces larger (see Table 2.3). This growth follows the progressive evolution of the sticking contact with μ_{loc}. The variation of the feed force F_f is more significant than for the cutting force F_c. When μ_{loc} varies from 0.2 to 1, F_f increases from 160 to 360 N, while F_c is about 500 N (see Table 2.3 and Figure 2.10). These simulation results also show the difference between the local friction coefficient μ_{loc} (in the sliding contact) and the apparent friction coefficient μ_{app}, which results from the whole contact between the cutting tool and the chip. Since the tool rake angle is zero, μ_{app} is given by $\mu_{app} = F_f/F_c$.

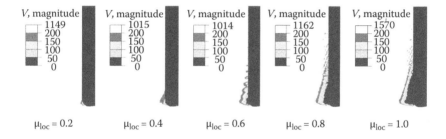

Figure 2.9 Effect of the local friction coefficient on the work material flow along the tool rake face.

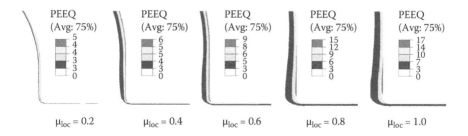

Figure 2.10 Equivalent plastic strain (PEEQ): influence of the local friction coefficient.

Table 2.3 Effect of the local friction coefficient μ_{loc} on cutting forces and apparent friction coefficient μ_{app}

Local friction coefficient			Apparent friction coefficient
μ_{loc}	F_c	F_f	μ_{app}
0.2	500	160	0.32
0.4	525	250	0.47
0.6	525	300	0.57
0.8	540	360	0.66
1.0	500	360	0.72

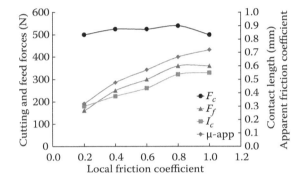

Figure 2.11 Effect of the local friction coefficient on cutting forces, apparent friction coefficient, and contact length.

From Table 2.3 or Figure 2.11, it appears that μ_{app} is an increasing function of μ_{loc}. Note that μ_{app} can be very close to μ_{loc} if the tool cutting edge radius is small enough and the tool–chip interface is a sliding contact. Indeed, the formation of the BUE and the sticking zone make the tribological conditions dependent on the thermomechanical conditions along the tool rake face and the rounded cutting edge.

2.4   Conclusion

The present work was focused on the coupling between the tribological conditions at the tool–chip interface and the BUE formation in dry machining of the aeronautical aluminium alloy A2024-T351. The chip formation and BUE processes strongly depend on the cutting conditions through the thermomechanical conditions at the tool–chip contact. Experimental results reveal that the influences, on the chip shape, of the cutting speed and the tool rake angle are negligible when the feed rate is small. In this case, we have continuous chips. In contrast, chip morphology significantly varies with the tool rake angle for large feeds.

To study the coupling between the thermomechanical behavior of the work material and the tribological conditions at the tool–chip contact, the model must be able to predict accurately the evolution of several contact parameters as temperature, plastic deformation, pressure, sticking, etc. To achieve this purpose, an ALE-FE model was developed by using the Abaqus/Explicit software. The numerical results reveal that the sticking contact increases with the local friction coefficient, inducing a BUE formation. The study presented in this work also shows the following:

- The ALE-FE approach is quite useful for examining the thermomechanical material flow around the rounded cutting edge and along the tool rake face even for large values of friction coefficient.
- BUE formation can result from a progressive or a sudden increase in friction coefficient when the chip interacts with the tool.
- For a rake angle of 0°, the influence of the local friction coefficient is more pronounced for the feed force than for the cutting force.
- The BUE corresponds to a work material sticking on the tool rake face resulting from the material thermal softening becomes more significant with large local friction coefficient μ_{loc}.
- The apparent friction coefficient that results from the whole contact between the cutting tool and the chip is an increasing function of μ_{loc}.

References

Abaqus V 6.9. Dassault Systems Simulia 2009.

Frang N, SrinivasaPai P, Mosquea S. The effect of built-up edge on the cutting vibrations in machining 2024-T351 aluminium alloy. *The International Journal of Advanced Manufacturing Technology* 2010;49:63–71.

Gomez-Parra A, Alvarez-Alcon M, Salguero J, Batista M, Marcos M. Analysis of the evolution of the built-up edge and built-up layer formation mechanisms in the dry turning of aeronautical aluminium alloys. *Wear* 2013;302:1209–1218.

Haddag B, Atlati S, Nouari M, Znasni M, Finite element formulation effect in three-dimensional modelling of a chip formation during machining. *International Journal of Material Forming* 2010;3(Suppl 1):527–530.

Haddag B, Nouari M, Barlier C, Dhers J. Experimental and numerical analyses of the tool wear in rough turning of large dimensions components of nuclear power plants. *Wear* 2014a;312:40–50.

Haddag B, Nouari M, Barlier C, Dhers J. Tribological behaviour and tool wear analyses in rough turning of large-scale parts of nuclear power plants using grooved coated insert. *Tribology International* 2014b;80:58–70.

Iwata K, Ueda K. Fundamental analysis of the mechanism of built-up edge formation based on direct scanning electron microscope observation. *Wear* 1980;60:329–337.

Kone F, Czarnota C, Haddag B, Nouari M. Finite element modelling of the thermo-mechanical behavior of coatings under extreme contact loading in dry machining. *Surface and Coatings Technology* 2011;205:3559–3566.

List G, Nouari M, Géhin D, Gomez S, Manaud JP, Le Petitcorps Y, Girot F. Wear behaviour of cemented carbide tools in dry machining of aluminium alloy. *Wear* 2005;259:1177–1189.

Mabrouki T, Courbon C, Zhang Y, Rech J, Nélias D, Asad M, Hamdi H, Belhadi S, Salvatore F. Some insights on the modelling of chip formation and its morphology during metal cutting operations. *C. R. Mecanique* 2016;344:335–354.

Nouari M, List G, Girot F, Géhin D. Effect of machining parameters and coating on wear mechanisms in dry drilling of aluminium alloys. *International Journal of Machine Tools & Manufacture* 2005;45:1436–1442.

Ohgo K. The adhesion mechanism of the built-up edge and the layer on the rake face of a cutting tool. *Wear* 1978;51:117–126.

Rahim EA, Sasahara H. A study of the effect of palm oil as MQL lubricant on high speed drilling of titanium alloys. *Tribology International* 2011;44:309–317.

Ramaswami R. The effect of the built-up edge (BUE) on the wear of cutting tools. *Wear* 1971;18:1–10.

Selvam MS, Radhakrishnan V. Groove wear, built-up edge and surface roughness in turning, *Wear* 1974;30(2):179–188.

Senthil Kumar A, Raja Durai A, Sornakumar T. Wear behaviour of alumina based ceramic cutting tools on machining steels. *Tribology International* 2006;39:191–197.

Sukvitfayawong S, Inasaki I. Detection of built-up edge in turning process. *International Journal of Machine Tools and Manufacture* 1994;34:829–840.

chapter three

Advances in the machining of holes and internal threads in light alloys

Carlos Henrique Lauro, Lincoln Cardoso Brandão,
Diego Carou, and J. Paulo Davim

Contents

3.1 Introduction

Although drilling, reaming, and tapping appear to be simple operations, they can have high complexity and cost involved. Drilling, reaming, tapping, and counter boring operations are predominant cycle-time-defining operations because they are carried out after a lot of time and money have already been invested in the parts. Thus, these operations require great accuracy (shape and machined surface), mainly when they are used for assembling parts.

The great success of assembling techniques in industry, such as joining, can be justified by the usage of lightweight materials (aluminum and magnesium) (Gröber et al. 2015). However, tool failure, such as the adhesion of material, represents an expensive reworking or the rejection of workpieces (Steininger et al. 2015). To avoid scrap or reworking, the tools should offer high process reliability (Tönshoff et al. 1994). Tapping processes, which are generally one of the last operations, can represent significant added costs when the taps fail (Bhowmick et al. 2010a). Moreover, the difficulty in directing the fluid flux to the cutting zone in processes such as tapping, deep-hole drilling, and milling can cause breakage of the tool (Sugihara and Enomoto 2009).

In drilling, burr formation can occur frequently. In drilling of metal sheets, plastic deformation can cause burr formation on the entry and the exit of the hole (Pilný et al. 2012). The removal of burrs can be a great part of the cost in the manufacturing of several components, such as automotive parts (15% to 20%) or aircraft engines (30%) (Dornfeld and Min 2010). Thus, it is important to develop methods to minimize or prevent burr formation, which require good knowledge of the design of the component.

Based on the previous information, this chapter discusses the importance of drilling, reaming, and tapping processes in the machining of light alloys. Light alloys attract great industrial interest; however, these materials require considerable attention during the machining due to, among others, the high build-up-edge (BUE) formation, low ignition point, or low thermal conductivity, depending on the material. Although these operations, mainly drilling, are widely used, there is a need to discuss non-traditional drilling techniques, drilling in sheets, type of taps, the use of cutting fluids, and the imperfections and faults, such as burrs, that can damage the part or its working capacity.

3.2   Drilling

Drilling is a process that creates or enlarges a hole in a workpiece. Generally, it can be considered as an intermediate process (comparing milling and turning) because the drill has usually two cutting edges that work continuously. This process is characterized by the variation of cutting speed in the cutting edge, which increases from nearly zero, in the center, to the work speed, in the periphery of the drill (Childs et al. 2000).

According to Trent and Wright (2000), the features of this process can be caused by the variations in cutting speed and rake angle. The authors highlighted that the tool should be slender and highly stressed, and the flutes required meticulous design to keep adequate strength during the chip flow. Besides, the rake face in the twist drill is constituted by part of each of the flutes and the rake angle is defined by the helix angle, which is defined for specific types of material.

According to Balajia et al. (2018), during drilling, the cutting and chip formation cannot be visualized, making it difficult to control vibrations that cause poor surface quality, high noise, and damage of the tool and/ or machine. These authors analyzed the influence of spindle speed, helix angle, and feed rate during drilling of Ti-6Al-4V titanium alloy. The authors observed that the helix angle had a lower influence than did the spindle and feed speeds on surface roughness. However, the use of high spindle speeds and helix angle increased surface roughness due to the rapid tool wear. The vibration of the tool was reduced when the helix angle was decreased, and the influence of the feed speed was not very important.

Silva et al. (2015) analyzed the hole size, roundness, radial and diameter deviation, and roughness in the drilling of SAE 323 aluminum alloy. The authors used a tool with diamond-like carbon (DLC) coating, minimum quantity lubrication (MQL) system, cutting speed of 217 m/min (11,500 rpm), and feed rate of 0.25 mm/rev. This coating can be a great option for drilling aluminum alloys at high speeds. This is because, when compared with the uncoated tools, an improvement in the roundness; deviations in diameter, roughness, and hardness; and reduced friction were observed.

In drilling of AM60B magnesium alloy, Gariboldi (2003) compared uncoated to different physical vapor deposition (PVD) coated drills in terms of tool life, wear, and hole finish. Drilling tests were developed at constant cutting speed, 63 m/min, and different feed rate. The authors observed that the feed rate influenced the level of material adhesion on the tool/drilled surface. When analyzing the use of low feed rates (0.27 and 0.37 mm/rev), the titanium nitride (TiN) coating presented great results of tool life. But for the higher feed rates, the chromium nitride (CrN) and zirconium nitride (ZrNb) coatings provided the best results.

3.2.1   Friction drilling

Friction (thermal, flow, or friction stir) drilling is a potential technique to machine holes that uses a special drill (Figure 3.1). This process consists of four stages. First, the contact between the center region and the workpiece produces friction due to the axial force and the tool rotation that causes heat and softens the workpiece. As the conical region enters the workpiece, the softened workpiece is pushed sideward and upward. Then, the conical region can pierce through the heated and softened workpiece. To finish the process, the cylindrical region is displaced to push the material and form the bushing. The hole and bushing formation is important in the initial stage of the process to extend its use in the industry (Dehghan et al. 2017).

According to Miller et al. (2006), an important parameter in friction drilling is the workpiece thickness/drill diameter ratio (t/d). When this ratio is high, it indicates that a larger portion of the material is displaced

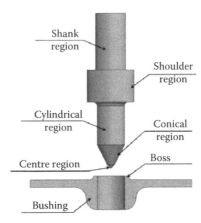

Figure 3.1 Friction drill.

and contributes to the bushing forming. In addition, an important characteristic of the material is the melting temperature. This process is relatively new, and most studies focus on the tool geometry and the mechanical performance, but some advantages can be identified compared to conventional drilling, such as the absence of waste material or chips, rapidity, low cost, and higher tool life (Eliseev et al. 2017).

To study the friction drilling of Al6061 aluminum alloy, Dehghan et al. (2017) developed a thermomechanical finite element model with experimental validation. The authors observed that the temperature during drilling can reach the melting temperature. The temperature in the workpiece in the initial stage was around 90°C, which increased close to 330°C along the contact area in the second stage and observed a peak around 650°C in the third stage. In the last stage, the temperature decreased, being uniformly distributed in the hole-wall due to the contact between the cylindrical part and the workpiece.

Miller et al. (2006) analyzed friction drilling (t/d = 0.75) in 4.0-mm-thick sheets of Al380 aluminum alloy and AZ91D magnesium alloy, a lightweight magnesium-aluminum-zinc alloy. Preheating of the workpiece improved the results, reducing the thrust force, torque, energy, and power. For Al380 aluminum alloy, less severe cracking and petal formation on bushings occurred at the higher temperature, 300°C. However, for magnesium alloy, preheating was not employed due to the exothermic oxidation that can ignite inside the oven during heating.

Eliseev et al. (2017) investigated the alteration in the grain structure of the Al2024 aluminum alloy, analyzing the stir (SZ), thermomechanically affected (TMAZ), and heat affected (HAZ) zones (Figure 3.2). In the SZ,

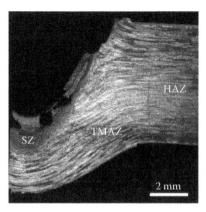

Figure 3.2 Cross-section of a friction drilled hole, interrupted operation. (Reprinted from *Mater Sci Eng A*, Eliseev AA, Fortuna SV, Kolubaev EA, Kalashnikova TA, Microstructure modification of 2024 aluminum alloy produced by friction drilling, Copyright 2017, with permission from Elsevier. License number: 4121081274232.)

with direct contact with the friction, heating, and deformation, grain structure is recrystallized due to high temperature and deformation, showing fine equiaxial grains. In the TMAZ, with deformation and high temperature without tool contact, the grains were large and elongated and oriented in the deformation direction, larger than the base material in both directions, indicating a growth due to high temperature. The grain structure in the HAZ was not examined.

In the study of the effects of temperature and plastic strain, Miller et al. (2005) used different materials: AISI 1020 and 4130 steel, Al 5052 aluminum alloy, and commercially pure titanium. When compared to steel and aluminum alloy, friction drilling in commercial pure titanium is a hard process, forming a bushing short and thick. The authors recommend drilling at low tool rotational speed and using lubricants to properly obtain hole and bushing, but the internal surface can present damages.

Biermann and Liu (2014) studied temperature using an infrared camera, thrust force, and torque in the drilling of AZ31 magnesium alloy. The authors used a wrought alloy with a thickness of 4 and 5 mm and tool diameters of 5.0 and 5.4 mm. The feed rate and wall thickness influenced the thrust force and torque. The thrust force and torque were low at the beginning due to low contact area. The thrust force increased before the conical region entered into the workpiece and increased again. The torque increased almost linearly during the process. Higher temperature was detected at the conical region of the tool into the bore wall because this region caused most of the deformation.

3.2.2   Imperfections and faults that occur in the drilling process

A wide range of burrs can be observed in machining operations. In drilling, the feed rate is the main parameter that leads to burr formation. However, the geometry and tool/work orientation can be of great influence for burr formation because the hole axis can be orthogonal or not to the plane of the exit surface of the hole (Dornfeld and Min 2010).

In drilling, exit burr, which is a rollover burr, occurs when the shear energy is larger than the energy for plastic deformation (Figure 3.3; Chang and Bone 2010). Burr is a big problem in drilled mechanical components that can reduce their quality, stunting the assembly, decreasing precision engineering, and increasing the cost (deburring). Analytical modelling, investigating the interaction of the tool/material, and varying cutting parameters are techniques used to understand the formation mechanism that can prevent or reduce burr formation (Kundu et al. 2014).

Abdelhafeez et al. (2015) studied the effect of cutting parameters on burr size, hole quality, and tool wear of Ti-6Al-4V titanium, Al2024 and Al7010 aluminum alloys. For the hole out-of-roundness and diameter oversize, the authors did not observe a relationship with cutting parameters because some aspects, such as vibration (fixture/machine), damping characteristics, and lateral tool deflection/vibration, also contributed to the highly nonlinear variation.

In the drilling of aluminum alloy sheets with a thickness of 2 mm, Pilný et al. (2012) analyzed the thrust force and burr formation using

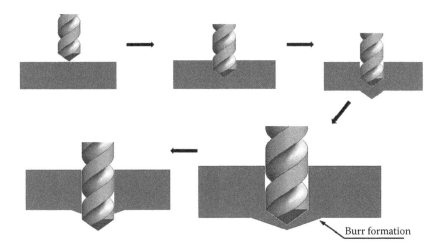

Figure 3.3 Drilling burr formation process. (Reprinted from *Precis Eng*, 34, Chang SSF, Bone GM, Burr height model for vibration assisted drilling of aluminum 6061-T6, 369–375, Copyright 2010, with permission from Elsevier. License number: 4123340729685.)

different drill geometries of 2 mm. The drill with double cone (chamfered) point geometry provided excessive entry burr heights larger than other drills (2 to 3 times). The 3-flute drill D required lower thrust force than other drills did (e.g., 2-flute), which caused a lesser plastic deformation, causing a smaller or burr-free edges at hole exit.

According to Bhowmick et al. (2010b), the use of MQL in the machining of aluminum alloys shows similar performance to the wet condition. However, there are gaps about the feasibility of MQL for magnesium alloys. In the drilling of AM60 magnesium alloy with high-speed steel (HSS) drills, the authors observed, in dry conditions, a prominent BUE formation due to high torques and thrust forces, a rise in temperature, and tool breaking before 80 holes. For use of the MQL, the drill reached the limit laid (150 holes), which presented no abrupt increase in torque and thrust force and small and discontinuous chips, and the machined surface was like wet condition.

During drilling, the use of cutting fluids can provide lubrication, cooling, and chip transport; however, it is complex and the efficiency of cutting fluids depends on fluid type, application method, workpiece and tool material, tool geometry, machining parameters, and drilling procedure (Jayal et al. 2007). Asok and Chockalingam (2014) compared the dry and compressed cold air drilling of Al6061 aluminum alloy. In general, the machined surfaces in the dry condition are as smooth as the machined surfaces in the cold air condition because the coating offers great durability and material removal rate that ensure hole quality. The accuracy of holes can be impaired due to the high temperatures that lead to the adhesion of chips on the tool in the dry condition or due to the tool wear in the cold air condition. For both conditions, the thrust force was lower using higher feed rates and spindle speeds.

Comparing the dry condition and different setups for MQL and wet condition during drilling of 10 mm in A390.0 cast aluminum alloy, Jayal et al. (2007) observed that for depths between 20 mm and 40 mm, the perpendicularity was less than 5 um/mm (3°). In the dry condition, dimensional accuracy was not dependent on cutting parameters, and it was lower than the dimensional accuracy in the MQL and wet conditions. The authors considered that the wet conditions were the method that offered higher accuracy.

3.2.3   Techniques to improve drilling

The surface finish, form accuracy, tool life, burr formation, and machining forces can be improved when employing vibration-assisted machining (VAM) (Brehl and Dow 2008). For instance, in drilling of Al6061-T6 aluminum alloy with a thickness of 3.18 mm, Chang and Bone (2010) employed TiN-coated drills and VAM, a frequency of 4 to 12 kHz, and an amplitude of 0.002 mm.

In drilling of UNS M11917 magnesium alloy, Berzosa et al. (2017) analyzed the influence of the point angle (118° and 135°) when employing dry and MQL conditions. The authors observed that the surface roughness (R_a) was not influenced by the point angle, although the point angle interactions with cutting speed and feed rate were significant. The MQL system reduced the surface roughness when compared with dry machining, being the optimal condition for the use of the MQL system.

Bhowmick and Alpas (2011) studied the drilling of AZ91 magnesium alloy using nonhydrogenated diamond-like carbon (NH-DLC)-coated tools using dry machining and the MQL condition. For the dry condition, the highest torques generated in the first tests that limited the tool life were at 66 (uncoated) and 25 (NH-DLC). The NH-DLC coating reduced the average coefficient of friction in the dry condition, from 0.39 to 0.27. However, when using the MQL system, the average coefficient of friction was reduced from 0.29 (uncoated) to 0.11 (NH-DLC). When comparing the dry cut to MQL system, the authors observed a decrease in the temperature from 175°C to 83°C and from 271°C to 52°C for uncoated and NH-DLC-coated tools, respectively.

Although the use of the MQL system can provide equal or even better results when comparing to wet condition, for processes like deep-hole drilling of titanium and its alloys, the MQL system has some gaps (Tai et al. 2014). In the deep hole drilling of EN AC-46000 aluminum cast alloy, Biermann et al. (2012) observed low process reliability and deviations in the workpiece dimensions due to the increase in the thermal load when replacing the emulsion by the MQL system. These thermal and mechanical loads were influenced by the feed rate; the lower the feed rate, the higher the thermal load.

The great use of titanium alloys due to their structural and corrosion-resistant characteristics encouraged Pujana et al. (2009) to study the ultrasonic vibration-assisted drilling in Ti-6Al-4V titanium alloy. The authors observed a reduction in feed force, mainly when the vibration amplitude was high. However, the temperature in the tool tip increased when ultrasonic vibration-assisted drilling was used.

3.3   Tapping process

Among the machining processes, threading can be considered as one of the most important processes for several industrial sectors such as automotive and aerospace. According to Dias et al. (2014), although threading is a complex process, because the breakage of the tool into the hole can occur, undermining the productivity of the process, about 50% of holes are threaded, representing the machining of holes 60% of all machining processes.

In modern machine tools, due to computer numerical controls (CNCs), threading manufacturing became conventional, which is important in order to meet the large number of internal threads required by sectors such as the automotive industry, which needs this process with stability and productivity (Steininger et al. 2015). Uzun and Korkut (2013) recommend the use of tap pulling heads and safe tap holders when the CNC machining center has no rigid tapping property.

According to Bratan et al. (2016), the working capacity of the threaded joints is determined by the internal thread quality (tool quality, conditions, parameters, coolants, and other), and the accuracy of the thread can be defined by means of accounting of processing errors (Equation 3.1):

$$\Delta_{D_2'}^{\Sigma} = \Delta_H D_2' + \Delta_T D_2' + \Delta_P D_2' + f_d^m + f_S^m + f_a^m + \Delta_a \qquad (3.1)$$

where $\Delta_H D_2'$ is the measurement error introduced by the control gage; $\Delta_T D_2'$ is the temperature error; $\Delta_P D_2'$ is the error of breakage of thread pitch diameter; f_d^m is the inaccuracy of tap thread pitch diameter; f_S^m is the inaccuracy of pitch of thread; f_a^m is the inaccuracy of angle of thread; and Δ_a is the error of cut thickness increment because of the BUE.

According to Uzun and Korkut (2013), the correct choice of tap and its parameters is of great importance because it affects the quality and cost in the threading process. In this sense, it is important to analyze the research done in the field, particularly for light alloys. So, in Figure 3.4, the number of papers about tapping found in the Engineering Village (2017) database is shown. When searching the terms "TAPPING" and "MACHINING" in all fields, the results show 41.8 papers per year, on average, between 1969 and 2017. However, when specifying a light alloy, "ALUMINUM and ALLOYS," or "MAGNESIUM and ALLOYS," or "TITANIUM and

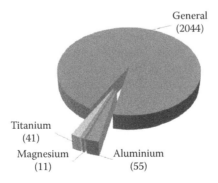

Figure 3.4 Number of paper about tapping process between 1969 and 2017. (Courtesy of Engineering Village, number of published papers about tapping process, available at www.engineeringvillage.com, 2017.)

Table 3.1 Repeatability, variability range, and resolution
of different cutting force tests

Process	Material	Cutting fluid	σ (%)	ρ (%)	σ/ρ
Tapping	Stainless steel	Water based	0.5–5	10	0.2
Tapping	Aluminum alloy	Water based	7–40	80	0.3
Reaming	Aluminum alloy	Water based	5–50	80	0.4
Drilling	Aluminum alloy	Oil	1–7	8	0.5

Source: De Chiffre, L., Belluco, W., *CIRP Ann. Manuf. Technol.*, 49, 57–60, 2000.

Note: σ is the experimental standard deviation of measurements. ρ is the range of variability of average measurements. σ/ρ is the indicator of the relative resolution of a given testing method.

ALLOYS," the sum of the papers between 1969 and 2017 is 107 papers. Although the search results are not highly reliable, it can be observed that the tapping process is widely employed for threading, but there are still many gaps when considering the tapping process of light alloys.

De Chiffre and Belluco (2000) compared the cutting fluid performance for different metal cutting processes. According to the authors, the cutting force tests are the most common tests for tapping, drilling, and reaming. When observing the results found, the complexity of the tapping in the aluminum alloy can be seen. Reaming presents modest repeatability of torque and thrust tests compared to tapping, with a lower relative resolution (0.3 to 0.4). Table 3.1 shows the repeatability, variability range, and resolution for cutting force tests.

3.3.1   Types of tapping processes

The threaded joint is a solution widely used because it provides an assembly with high strength, high rigidity, disassembling for maintenance, and recycling. The internal threads, or nuts, can be manufactured by two main processes, cut or form threading, that are defined considering the process aspects, specifications, and characteristics (Fromentin et al. 2005). M8x1.25 is the most commonly used dimension of thread in the manufacturing industry (Uzun and Korkut 2013).

The cutting tap cuts thin slices perpendicular to the axial direction, and its complex geometry causes variation in the chip load due to the tool/workpiece position (Cao and Sutherland 2002). A cutting tap can be divided into three zones of threads. The first zone is a couple of chamfered threads; the second zone has full threads playing a role in finishing threading and guiding the tool itself into the thread; and the third zone is composed of the relief angle guides to create threads when retreating in backward tapping (Ahn et al. 2003).

Hsu et al. (2016) studied the effects of geometric parameters (helix angle and the chamfer length) in the tapping of Al 6061 aluminum alloy (M3 cutting tap). The choice of the helix angle for tapping is related to the material, larger helix angle for the soft materials and smaller helix angle for the hard materials. However, a larger helix angle can be used for blind hole threads because it allows a better rate of expulsion of the material removed. The chamfer can influence the depth, the amount of cut, the force needed, and heat flux that can cause fatigue and damage. A longer chamfer reduces the cutting capabilities, and a shorter chamfer compromises durability and increases the workload.

Whereas in cut tapping, the thread is obtained by chip removal, in form tapping, the thread is formed only by the displacement of the material (Figure 3.5). Form threading is characterized by an incomplete thread profile, an appearance of a split crest on the top of the thread (Fromentin et al. 2005).

Generally, the initial hole can be defined by Equations 3.2 and 3.3 for cut and form tapping, respectively. The initial hole for form tapping is higher than for cut tapping. In cut tapping, the crests of the threads are built in the wall of initial hole, as to form tapping; the crests of the threads are formed by the displaced material (Figure 3.6):

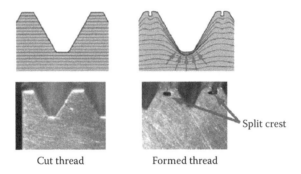

Cut thread   Formed thread

Figure 3.5 Thread profile in the Cut and forming tapping. (Based on Emuge Franken, InnoForm Cold-forming taps [tool catalogue], 2007.)

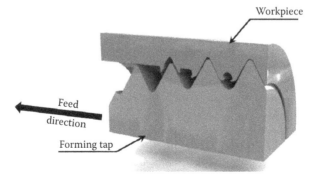

Figure 3.6 Formed thread formation. (Based on Emuge Franken, InnoForm Cold-forming taps [tool catalogue], 2007.)

$$D_H = D_T - p \qquad (3.2)$$

and

$$D_H = D_T - \frac{p}{2}, \qquad (3.3)$$

where D_H is the diameter of initial hole, D_T is the diameter of threading, and p is the pitch.

In conformed thread, the grain structure is strengthened in the root of the thread that is exposed to the mechanical efforts or crack formation, and this deformation increases the thread resistance (Emuge Franken 2007). Masmoudi et al. (2017) analyzed the split crest in tapping of 1000 class aluminum alloy. The results indicated that the height of the formed thread influences the split crest that influences the quality of threads. According to these authors, the quality can be obtained when analyzing the experimental height/theoretical height ratio, which, in their research, the authors found to be 66% and 71% for tapped drilled holes and tapped flanged holes, respectively.

When compared to cut tapping, a great advantage of form tapping can be the nonformation of chips and waste, which represents a reduction in the cleaning time and no need for waste disposal. A disadvantage is the increase in the power of the machine tool, which causes an increase in electric energy consumption. Thus, the choice of tapping processes should analyze the environment impact and electric energy costs, which can be minimized by the cleaning of the threads with spray blower in the last stage of manufacturing (Ribeiro Filho et al. 2017).

Wittke et al. (2015) compared different manufacturing techniques for internal threading in Al6060 aluminum alloy. When comparing the hardness of the profile obtained with the bulk hardness, these authors observed for cut tapping and mill threading that the changes in the hardness of the profiles were considered negligible. However, the crests and roots showed hardening of about 20% in bulk material in form threading and about 40% in flat profiles, in contrast to the basic hardness of the material. The reason was that the formed threads showed the best mechanical properties, whereas the mechanical properties for cut threads were poor because the threads were cut unevenly.

Carvalho et al. (2012) studied the form tapping process (M10) of AM60 magnesium alloy with hardness between 65 and 74 HB. The diameter of the initial hole and the forming speed were varied, using values above or below the recommended, and the tools used included coated and uncoated solutions. The authors observed that forming speeds exceeding 100 m/min caused the locking of the spindle that can influence the process finishing for this alloy.

Fromentin et al. (2010) highlighted the importance of the understanding of the frictional conditions to correctly define the fluids as coolants or lubricants, mainly in thread forming, because tapping is a very harsh machining process that is quite sensitive to lubrication or cooling. According to Hsu et al. (2016), the use of the cutting fluid decreases the wear on machinery during tapping processes due to increasing the lubrication between the machinery and parts.

According to Soković and Mijanović (2001), Flemming and Sudholz, in 1956, were the pioneers evaluating the performance of cutting oils in tapping using statistical approaches. This is relevant because in tapping tests, the use of cutting fluids with great performance characteristics is recommended. The authors affirmed that the mixture of alkiestres of phosphorus acid with condensation products of high grease acid, as cutting fluid, can be used as the first choice for aluminum and other nonferrous metals in specific procedures, such as tapping.

Cao et al. (1997) analyzed the R_a surface roughness in the cut tapping of Al390 and Al308 aluminum alloys (HSS tap ¼-20 UNC). When the spindle speed was reduced, the values of surface roughness decreased for Al308 aluminum alloy, but for Al390 aluminum alloy, these values increased. When comparing dry and wet conditions, the use of cutting fluids tended to reduce the surface roughness. The material was the most significant variable, which can be justified by the imperfections (porosity, smearing, and tearing off) and scratches by trapped chips.

Hong et al. (1993) were concerned with the influence of the different overbased sulfonates in the tapping of the Ti-6Al-4V titanium alloy, 2C ¼-20 NC. The authors observed, in two sulfonates types, the appearance of a smooth surface. Nevertheless, the use of overbased magnesium sulfonate caused plastic deformation, indicating an adhesive wear. Ribeiro Filho et al. (2017) analyzed the behavior in internal threading (M10) for Al306 aluminum alloy with different lubrication/cooling systems: dry, wet, and three specific vegetable fluids for MQL systems. When analyzing torque in form tapping, the authors observed similar values for one vegetable fluid to the use of emulsion. However, for cut tapping, two vegetable fluids provided a reduction (lower than 10%) in the torque values compared to wet condition.

According to Derflinger et al. (1999), tool life can be increased if the tool is coated by a combination of a hard and lubricant layer, like TiAlN (hard) and WC/C (lubricant) layers, which help in reducing the amounts of cutting fluids used, working with minimal or no lubrication in tapping processes.

Ribeiro Filho et al. (2016) compared cut and form tapping (M3) for the internal threading of the Ti-6Al-4V titanium alloy. Among different comparisons, the authors developed corrosion tests in synthetic plasma solution at a temperature of 39°C to analyze the biocompatibility of internal threads. A better biocompatibility was obtained with machining tapping at a speed of 2 m/min. In machining of thread holes (M3 to M6) in Al5052 aluminum alloy, Bratan et al. (2016) observed that the use of combined deforming-cutting taps can improve the accuracy and surface quality of threads. The tolerance limits, for grade 4H, were between 45% (forming-cutting taps) and 75% (forming tap).

In the threading (M6) of AZ31 magnesium alloy with thin walls, Biermann and Liu (2014) analyzed cut and form threading processes. The temperature in form threading was higher than in cut threading, 55°C and 40°C, respectively. In the forming process, microfractures due to insufficient formability at low temperatures were observed.

In the study of the tapping of Al396 (three chemical compositions), B319.2, and Al356.2 aluminum silicon alloy, Zedan et al. (2010) used an M8 HSS cutting tap with TiN coating. The experiments were carried out in a wet condition, using an emulsion of 5% cutting fluid and 95% liquid to avoid the influence of heat during the cutting. The authors observed that increasing iron content with the same level of Mn was most effective for the cutting forces and tool life because the detrimental effect of Fe may be partially neutralized by adding Fe in proportions greater than double the Mn concentration.

Dias et al. (2014) highlighted the great results obtained in form threading in magnesium, in industrial applications such as gearbox and engine heads for the automotive sector and, in aluminum, in both aeronautic and automotive applications. The authors studied the helical form tapping (M12) in AM60 magnesium alloy with a cobalt HSS (Figure 3.7), which has an advantage because of its high productivity. The reason is that the tool does not lock because the initial diameter is larger than the tool diameter. Great results, for torque and surface finishing, were obtained with a high forming speed of 150 m/min.

Sales et al. (2009) analyzed the use of tap-mill tool in ISO Al-Si-Cu4 aluminum alloy. The authors considered this technique as a promising alternative to manufacturing internal threads without the previous hole, eliminating the need for tool or machine change, which they used to machine different threaded holes (M6 or M7). The tap-mill begins the process as a drill, making the chamfered hole, blind or through. To finish the hole, the tool returns one third of the pitch. The next step is the displacement of the tool center line, approaching the hole wall to start the rotation and translation movements that milled the ridges. After this step, the operation is inversed; the tool returns to the center line and sets back to the original position.

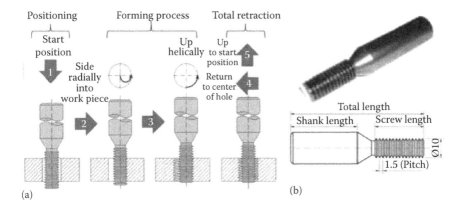

Figure 3.7 Movement (a) and geometry (b) of tool in the helical form tapping. (From Dias, L.D., Brandão, L.C., Ribeiro Filho, S.L.M., Coelho, R.T., *Mater. Manuf. Process*, 29, 748–753, 2014. With permission.)

3.3.2 Imperfections and faults occurring in the tapping process

With the growing need for bolts and threads, even down to 1 mm in diameter, to assemble or to fasten of components on notebook computers, hard disks, mobile phones, and others, machining should be done using high-speed spindles to avoid BUEs, and high acceleration/deceleration characteristics because of the short tapping length (Ahn et al. 2003). For difficult-to-cut materials, for instance, titanium alloys, tapping smaller than M6, mainly for deep holes, is a hard operation because of the common breakage of taps that occur due to galling and seizing (Zhang and Chen 1998).

Aluminum alloys are materials with high anticorrosion properties, relatively high strength, low weight, and high electrical conductivity and they can be easily cast, rolled, forged, stamped, and machined. However, aluminum alloys present significant difficulties in machining operations because these materials have high viscosity and tendency to adhesive bonding, which cause BUE on the tool cutting edge. For small internal threads (M3 to M6), BUE occurs on the rake face of the cutting tool, which can have a positive or negative effect. BUE reduces the cutting forces and tool wear (partial protection of tool working surfaces), but BUE changes the tool geometry and degrades surface roughness (Bratan et al. 2016).

Bhowmick et al. (2010a) studied the effect of MQL in the threading (cut tapping M8) of Al319 aluminum-silicon alloy. For the dry condition, the uncoated taps showed material adhered on the pitches, covering almost the entire area that caused the BUE, and in the coated tap, some patches were observed, but to a lesser extent. In the MQL and wet conditions, the authors did not observe material adhered on taps (no BUE formation).

Ribeiro Filho et al. (2015) studied the burr formation in form tapping of Al 7075 aluminum alloy using M10 tool in both coated (TiN-PVD) and uncoated options. To evaluate burr formation, the mean burr height of four equidistant points at 90° at the entrance and exit of the tapping was calculated. The authors observed that the burr formation was independent of the tools, diameter, or forming speed, being the mean height values between 0.051 and 0.842 mm. The variation of the initial diameter affected only the burr formation at the entrance. The exit of the tool was mostly affected by the forming speed. On the other hand, the burr formation was not influenced by the type of tool (coated or uncoated).

In the form tapping of class 1000 aluminum alloy (sheet of 2 mm), Masmoudi et al. (2017) analyzed burr formation at the entrance and exit. The measured heights varied between 0.1 and 0.22 mm. The smaller burrs occurred at the entrance of the flanged hole due to the presence of the curved profile that was equal to 0.33 mm.

To investigate the feed error caused by the servo system, Wan et al. (2017) used Al7050 aluminum alloy in the tapping process because the feed error can often cause excessive load, breakage of the tap, and damage of the machined parts. Analyzing the total tapping forces, the authors observed that the axial force, F_Z, was growing during the penetration of the cylindrical part because, although the threads on the tap's cylindrical part do not produce material removal, the feed error produces an indentation effect for every engaged thread of the cylindrical part.

Mezentsev et al. (2002) studied the fault detection in cut tapping (M10) in Al356 aluminum alloy using the radial forces. Based on the observation of cut signals, it was affirmed that:

- The radial forces at steady-state cutting are "nonzero" values when the axis is misaligned. The steady-state average torque remains the same as well as for the no-faults condition.
- A rise in the radial forces at steady-state cutting, keeping the average equal to zero, indicates tooth breakage, which does not affect the torque at steady state.
- A significant rise in the radial force magnitude, while their mean stays at zero, represents tap runout, and in the steady-state torque, a slight rise occurs.

Wittke et al. (2015) raised the awareness of the influence of drilling in internal threading. The friction drilling process caused oval forms of the holes that tolerate lower maximum loads. To study the machinability of new Al-Cu casting alloys, Elgallad et al. (2010) chose drilling and cut tapping (M8) processes, analyzing the cutting force and moment, tool life, and chip characteristics. To characterize the hole accuracy, the Go/No-Go test was used, observing deterioration after up to 1620 or 1800 holes due to the tooth-breakage. The authors highlighted that the precision of dimensions, shape, and surface finish are important assessment criterion to analyze the quality of hole.

Carvalho et al. (2012) analyzed the influence of the diameter in the form tapping in AM60 magnesium alloy (M10). Although the reduction of the hole diameter caused an increase in the thrust force and torque, in general, the fill rate also increased, mainly for coated tools. However, when comparing the hardness, the values increased or decreased, depending on where the measurement was performed.

3.3.3   Techniques to improve internal threading

Specifically, in the tapping process, Patil et al. (1987) developed a vibratory tapping attachment to obtain overload safety against torque; compensation for the error between the feed and pitch of the tap; cushioning against tap bottoming; parallel floating; and superimposition of torsional vibrations on cutting velocity with the facility to vary the amplitude and frequency independently. The authors observed a reduction in thrust force (56%) and torque (14%) in the tapping (M10) of aluminum alloy.

Zhang and Chen (1998) analyzed the influence of assisted vibration in the relief face friction during the tapping of two types of titanium alloys, Ti22 and Ti-6Al-4V (M3.5 cutting tap). The authors classified the mechanisms for the machined surface in vibration tapping into the three types:

- A rubbing process in which the spring-back and the frictional stress have a direct dependence on cutting depth.
- The ploughing process, in which the frictional stress is constant.
- The cutting process, in which the spring-back and the frictional stress are constant.

Pawar and Joshi (2016) analyzed the effects of VAM in the cut tapping of Ti-6Al-4V titanium alloy (M3 and M4). When comparing the conventional tapping with axial vibration, a reduction in the maximum torque (169.9 N·cm to 119.3 N·cm) and temperature (101°C to 80.5°C) was observed. However, the reduction was more important when employing the axial-torsional vibration for the maximum torque (93.2 N·cm) and temperature (69.7°C). It occurs because, in axial vibration, the relief faces are detached from the grooves, which allows the chips slide over the rake faces. In the axial-torsional vibration, the tool is detached from chips, reducing the heat conduction (tool, workpiece, and chips), i.e., a convective cooling throughout the process.

To study the quality of the formation of the profile, Tsao and Kuo (2012) studied tapping in Ti-6Al-4V titanium alloy, using three kinds of cutting taps (M3.5 with different coatings) and two techniques for ultrasonic-assisted vibration, fixed frequency (FF) and limited tracing frequency (LTF). Both ultrasonic-assisted vibrations substantially improved the quality of the thread area, regardless of coatings, when compared to conventional tapping. The authors identified a rise of 26.0%, on average, for the TiN coating and the FF-type ultrasonic-assisted vibration and 30.2% for the LTF-type and TiAlN Micran coated.

Kuo (2007) analyzed the influence of the cutting fluid in vibration-assisted tapping (M3 and M3.5) in commercially pure titanium. According to the authors, the smaller internal threading in pure titanium metal can be improved when using vibration-assisted tapping, reducing the tapping torque and risk of tap breakage and having no adverse effect on the profile. The use of the cutting fluid significantly reduced the tapping torque. However, the use of vibration-assisted tapping caused a higher reduction.

In the internal threading (M8) of Ti-6Al-4V titanium alloy, Uzun and Korkut (2013) analyzed the influence of cryogenic treatment on tap. The authors used different taps: TiAlN coated, uncoated, and cryogenically heat treated in tests under both dry and wet conditions. The tap with the cryogenic heat treatment produced lower reverse torques when compared to the uncoated and coated tools. In dry conditions, the cutting torque was also lower for cryogenic heat-treated taps than for the coated and uncoated tools.

The efficiency of the water-soluble bismuth-containing compound in the tapping of Al2024 aluminum alloy was investigated by Fang et al. (2010). The tests were developed with an HSS tool for M10 threading. The authors observed that a water-soluble organo–bismuth complex improved efficiency in the tapping of this alloy. Scanning electron microscopy indicated an excellent quality of the machined surface; nevertheless, a chemical reaction film (bismuth sulphide, ferrous sulphide, ferric oxide, ferrous sulphate, and iron phosphate) appeared on the surface.

3.4   Other processes

Similar to friction drilling, flow drill screwdriving (FDS) uses softening by frictional heat to form an extrusion with a rotating conical tool. However, in friction drilling, the process finishes after the extrusion, whereas in FDS, tapping the extrusion and installing a fastener are required; i.e., it saves production time because it combines several steps into a single operation, which is critical for the automotive sector (Milner et al. 2016).

3.5   Conclusions

The facility to assemble/disassemble equipment for maintenance operations makes threading the most popular joint technique. For the use of nuts and screws, drilling requires great quality in terms of dimensions and a good finishing process to avoid loosening, breakage, and damage in the profile as the equipment is assembled, among others. Among the threading processes, the importance of the tapping process in several industries sectors is clear, being a simple operation, manual, or automated process. However, it is a process of great aggregated value. The use of these processes, drilling and tapping for light alloys, needs to fill up many gaps, given that these materials draw attention to the automobile industry, for example, because of their resistance/weight ratio. The replacement of ferrous alloys by aluminum and its alloys reduces the weight of vehicles, which reduces the consumption of fuel; i.e., a vehicle will be more efficient and ecologic. In addition, a large part of aluminum used in the world is recycled. Moreover, the social importance of these materials, such as the use of titanium and its alloys in biocomponents, should be considered.

The importance of these operations and light alloys encourages carrying out more research in drilling and tapping used for these alloys. The development of studies about residual stresses in titanium alloys to reduce the biocorrosion in implants or the avoidance of build-up edge in aluminum alloys that can damage the threading is of great importance. The use of finite element method, with experimental validation, could help to fill up these gaps. Furthermore, the development of cheaper techniques for tool condition monitoring, like the electric current monitoring, can reduce waste during the production.

Acknowledgment

The authors gratefully acknowledge the Ministry of Education's Coordination for the Improvement of Higher Education Personnel (CAPES) and the Emuge Franken—Brazil. The authors would like to thank Elsevier and Taylor & Francis for granting permission for reusing of the published materials.

References

Abdelhafeez AM, Soo SL, Aspinwall DK et al. (2015) Burr formation and hole quality when drilling titanium and aluminium alloys. *Procedia CIRP* 37:230–235. doi: 10.1016/j.procir.2015.08.019

Ahn JH, Lee DJ, Kim SH et al. (2003) Effects of synchronizing errors on cutting performance in the ultra-high-speed tapping. *CIRP Ann Manuf Technol* 52:53–56. doi: 10.1016/S0007-8506(07)60529-0

Asok P, Chockalingam P (2014) Dry and compressed air cooling comparative study on 6061 aluminium alloy drilling using coated drill. *Adv Mater Res* 903:45–50. doi: 10.4028/www.scientific.net/AMR.903.45

Balajia M, Venkata Rao K, Mohan Rao N, Murthy BSN (2018) Optimization of drilling parameters for drilling of TI-6Al-4V based on surface roughness, flank wear and drill vibration. *Measurement* 114:332–339. doi: 10.1016/J .MEASUREMENT.2017.09.051

Berzosa F, de Agustina B, Rubio EM (2017) Tool selection in drilling of magnesium UNSM11917 pieces under dry and MQL conditions based on surface roughness. *Procedia Eng* 184:117–127. doi: 10.1016/J.PROENG.2017.04.076

Bhowmick S, Alpas AT (2011) The role of diamond-like carbon coated drills on minimum quantity lubrication drilling of magnesium alloys. *Surf Coatings Technol* 205:5302–5311. doi: 10.1016/J.SURFCOAT.2011.05.037

Bhowmick S, Lukitsch MJ, Alpas AT (2010a) Tapping of Al–Si alloys with diamond-like carbon coated tools and minimum quantity lubrication. *J Mater Process Technol* 210:2142–2153. doi: 10.1016/j.jmatprotec.2010.07.032

Bhowmick S, Lukitsch MJ, Alpas AT (2010b) Dry and minimum quantity lubrication drilling of cast magnesium alloy (AM60). *Int J Mach Tools Manuf* 50:444–457. doi: 10.1016/j.ijmachtools.2010.02.001

Biermann D, Liu Y (2014) Innovative flow drilling on magnesium wrought alloy AZ31. *Procedia CIRP* 18:209–214. doi: 10.1016/j.procir.2014.06.133

Biermann D, Iovkov I, Blumb H et al. (2012) Thermal aspects in deep hole drilling of aluminium cast alloy using twist drills and MQL. *Procedia CIRP* 3:245–250. doi: 10.1016/J.PROCIR.2012.07.043

Bratan S, Novikov P, Roshchupkin S (2016) Application of combined taps for increasing the shaping accuracy of the internal threads in aluminium alloys. *Procedia Eng* 150:802–808. doi: 10.1016/j.proeng.2016.07.115

Brehl DE, Dow TA (2008) Review of vibration-assisted machining. *Precis Eng* 32:153–172. doi: 10.1016/j.precisioneng.2007.08.003

Cao T, Sutherland JW (2002) Investigation of thread tapping load characteristics through mechanistics modeling and experimentation. *Int J Mach Tools Manuf* 42:1527–1538. doi: 10.1016/S0890-6955(02)00108-6

Cao T, Batzer SA, Sutherland JW (1997) Experimental investigation of tapped thread surface roughness for cast aluminum alloys. *Manuf Eng Div* 6:189–195.

Carvalho AO, Brandão LC, Panzera TH, Lauro CH (2012) Analysis of form threads using fluteless taps in cast magnesium alloy (AM60). *J Mater Process Technol* 212:1753–1760. doi: 10.1016/j.jmatprotec.2012.03.018

Chang SSF, Bone GM (2010) Burr height model for vibration assisted drilling of aluminum 6061-T6. *Precis Eng* 34:369–375. doi: 10.1016/j.precisioneng.2009.09.002

Childs THC, Maekawa K, Obikawa T, Yamane Y (2000) *Metal Machining Theory and Applications.* Arnold, London.

De Chiffre L, Belluco W (2000) Comparison of methods for cutting fluid performance testing. *CIRP Ann Manuf Technol* 49:57–60. doi: 10.1016/S0007-8506(07)62895-9

Dehghan S, Ismail MIS, Ariffin MKA et al. (2017) Numerical simulation on friction drilling of aluminum alloy. *Materwiss Werksttech* 48:241–248. doi: 10.1002/mawe.201600768

Derflinger V, Brändle H, Zimmermann H (1999) New hard/lubricant coating for dry machining. *Surf Coatings Technol* 113:286–292. doi: 10.1016/S0257-8972(99)00004-3

Dias LD, Brandão LC, Ribeiro Filho SLM, Coelho RT (2014) Processing of threads on a magnesium alloy using a special process. *Mater Manuf Process* 29:748–753. doi: 10.1080/10426914.2014.901533

Dornfeld D, Min S (2010) A review of burr formation in machining. In: Aurich JC, Dornfeld D (eds), *Burrs—Analysis, Control and Removal: Proceedings of the CIRP International Conference on Burrs.* Springer Berlin Heidelberg, Berlin, Heidelberg, pp 3–11. doi:10.1007/978-3-642-00568-8

Elgallad EM, Samuel FH, Samuel AM, Doty HW (2010) Machinability aspects of new Al-Cu alloys intended for automotive castings. *J Mater Process Technol* 210:1754–1766. doi: 10.1016/j.jmatprotec.2010.06.006

Eliseev AA, Fortuna SV, Kolubaev EA, Kalashnikova TA (2017) Microstructure modification of 2024 aluminum alloy produced by friction drilling. *Mater Sci Eng A* 691:121–125. doi: 10.1016/j.msea.2017.03.040

Emuge Franken (2007) InnoForm Cold-forming Taps (tool catalogue).

Engineering Village (2017) Number of published papers about tapping process. https://www.engineeringvillage.com. Accessed May, 31 2017.

Fang J, Xia Y, Liu W (2010) The tribological behavior of bismuth dithiophosphate as water - based additive in aluminium alloy tapping. *Ind Lubr Tribol* 62:327–331. doi: 10.1108/00368791011076218

Fromentin G, Poulachon G, Moisan A et al. (2005) Precision and surface integrity of threads obtained by form tapping. *CIRP Ann Manuf Technol* 54:519–522. doi: 10.1016/S0007-8506(07)60159-0

Fromentin G, Bierla A, Minfray C, Poulachon G (2010) An experimental study on the effects of lubrication in form tapping. *Tribol Int* 43:1726–1734. doi: 10.1016/j.triboint.2010.04.005

Gariboldi E (2003) Drilling a magnesium alloy using PVD coated twist drills. *J Mater Process Technol* 134:287–295. doi: 10.1016/S0924-0136(02)01111-1

Gröber D, Georgi W, Sieber M et al. (2015) The effect of anodising on the fatigue performance of self-tapping aluminium screws. *Int J Fatigue* 75:108–114. doi: 10.1016/j.ijfatigue.2015.02.008

Hong H, Riga AT, Gahoon JM, Scott CG (1993) Machinability of steels and titanium alloys under lubrication. *Wear* 162–164:34–39. doi: 10.1016/0043-1648 (93)90481-Z

Hsu C, Yeh S, Lee J (2016) Effect analysis and optimal combination of cutting conditions on the cutting torque of tapping processes using Taguchi methods. *2016 IEEE International Conference on Automation Science and Engineering (CASE)*, Fort Worth, TX, pp. 1215–1218. doi: 10.1109/COASE.2016.7743544

Jayal AD, Balaji AK, Sesek R et al. (2007) Machining performance and health effects of cutting fluid application in drilling of A390.0 cast aluminum alloy. *J Manuf Process* 9:137–146. doi: 10.1016/S1526-6125(07)70114-7

Kundu S, Das S, Saha PP (2014) Optimization of drilling parameters to minimize burr by providing back-up support on aluminium alloy. *Procedia Eng* 97:230–240. doi: 10.1016/j.proeng.2014.12.246

Kuo K-L (2007) Experimental investigation of ultrasonic vibration-assisted tapping. *J Mater Process Technol* 192:306–311. doi: 10.1016/j.jmatprotec.2007.04 .033

Masmoudi N, Soussi H, Krichen A (2017) Determination of an adequate geometry of the flanged hole to perform formed threads. *Int J Adv Manuf Technol* 92(1–4): 547–560. doi: 10.1007/s00170-017-0145-0

Mezentsev OA, Zhu R, DeVor RE et al. (2002) Use of radial forces for fault detection in tapping. *Int J Mach Tools Manuf* 42:479–488. doi: 10.1016/S0890 -6955(01)00139-0

Miller SF, Blau PJ, Shih AJ (2005) Microstructural alterations associated with friction drilling of steel, aluminum, and titanium. *J Mater Eng Perform* 14:647–653. doi: 10.1361/105994905X64558

Miller SF, Tao J, Shih AJ (2006) Friction drilling of cast metals. *Int J Mach Tools Manuf* 46:1526–1535. doi: 10.1016/j.ijmachtools.2005.09.003

Milner JL, Gnäupel-Herold T, Skovron JD (2016) Residual stresses in flow drill screw-driving of aluminum alloy sheets. In: *2016 11th International Manufacturing Science and Engineering Conference*, Conference Sponsors: Manufacturing Engineering Division, Virginia, 1:1–6. doi: 10.1115/MSEC2016-8823

Patil SS, Pande SS, Somasundaram S (1987) Some investigations on vibratory tapping process. *Int J Mach Tools Manuf* 27:343–350. doi: 10.1016/S0890-6955(87)80007-X

Pawar S, Joshi SS (2016) Experimental analysis of axial and torsional vibrations assisted tapping of titanium alloy. *J Manuf Process* 22:7–20. doi: 10.1016/j.jmapro.2016.01.006

Pilný L, De Chiffre L, Píška M, Villumsen MF (2012) Hole quality and burr reduction in drilling aluminium sheets. *CIRP J Manuf Sci Technol* 5:102–107. doi: 10.1016/j.cirpj.2012.03.005

Pujana J, River A, Celaya A, López de Lacalle LN (2009) Analysis of ultrasonic-assisted drilling of Ti6Al4V. *Int J Mach Tools Manuf* 49:500–508. doi: 10.1016/J.IJMACHTOOLS.2008.12.014

Ribeiro Filho SLM, Oliveira JA de, Arruda ÉM, Brandão LC (2015) Analysis of burr formation in form tapping in 7075 aluminum alloy. *Int J Adv Manuf Technol* 84:957–967. doi: 10.1007/s00170-015-7768-9

Ribeiro Filho SLM, Lauro CH, Bueno AHS, Brandão LC (2016) Effects of the dynamic tapping process on the biocompatibility of Ti-6Al-4V alloy in simulated human body environment. *Arab J Sci Eng* 41:4313–4326. doi: 10.1007/s13369-016-2089-3

Ribeiro Filho SLM, Vieira JT, de Oliveira JA et al. (2017) Comparison among different vegetable fluids used in minimum quantity lubrication systems in the tapping process of cast aluminum alloy. *J Clean Prod* 140:1255–1262. doi: 10.1016/j.jclepro.2016.10.032

Sales WF, Becker M, Gurgel AG, Júnior JL (2009) Dynamic behavior analysis of drill-threading process when machining AISI Al-Si-Cu4 alloy. *Int J Adv Manuf Technol* 42:873–882. doi: 10.1007/s00170-008-1658-3

Silva WM, Jesus LM, Carneiro JR et al. (2015) Performance of carbide tools coated with DLC in the drilling of SAE 323 aluminum alloy. *Surf Coatings Technol* 284:404–409. doi: 10.1016/J.SURFCOAT.2015.09.061

Soković M, Mijanović K (2001) Ecological aspects of the cutting fluids and its influence on quantifiable parameters of the cutting processes. *J Mater Process Technol* 109:181–189. doi: 10.1016/S0924-0136(00)00794-9

Steininger A, Siller A, Bleicher F (2015) Investigations regarding process stability aspects in thread tapping Al-Si alloys. *Procedia Eng* 100:1124–1132. doi: 10.1016/j.proeng.2015.01.475

Sugihara T, Enomoto T (2009) Development of a cutting tool with a nano/micro-textured surface—Improvement of anti-adhesive effect by considering the texture patterns. *Precis Eng* 33:425–429. doi: 10.1016/j.precisioneng.2008.11.004

Tai BL, Stephenson DA, Furness RJ, Shih AJ (2014) Minimum quantity lubrication (MQL) in automotive powertrain machining. *Procedia CIRP* 14:523–528. doi: 10.1016/J.PROCIR.2014.03.044

Tönshoff HK, Spintig W, König W, Neises A (1994) Machining of holes developments in drilling technology. *CIRP Ann Manuf Technol* 43:551–561. doi: 10.1016/S0007-8506(07)60501-0

Trent E, Wright P (2000) *Metal Cutting*, 4th ed. Butterworth–Heinemann, Woburn.

Tsao C, Kuo K (2012) Ultrasonic-assisted vibration tapping using taps with different coatings. *Trans Nonferrous Met Soc China* 22:s764–s768. doi: 10.1016/S1003-6326(12)61801-9

Uzun G, Korkut İ (2013) The effect of cryogenic treatment on tapping. *Int J Adv Manuf Technol* 67:857–864. doi: 10.1007/s00170-012-4529-x

Wan M, Ma Y-C, Feng J, Zhang W-H (2017) Mechanics of tapping process with emphasis on measurement of feed error and estimation of its induced indentation forces. *Int J Mach Tools Manuf* 114:8–20. doi: 10.1016/j.ijmachtools.2016.12.003

Wittke P, Liu Y, Biermann D, Walther F (2015) Influence of the production process on the deformation and fatigue performance of friction drilled internal threads in the aluminum alloy 6060. *Mater Test* 57:281–288. doi: 10.3139/120.110712

Zedan Y, Samuel FH, Samuel AM, Doty HW (2010) Effects of Fe intermetallics on the machinability of heat-treated Al-(7–11)% Si alloys. *J Mater Process Technol* 210:245–257. doi: 10.1016/j.jmatprotec.2009.09.007

Zhang D, Chen D-C (1998) Relief–face friction in vibration tapping. *Int J Mech Sci* 40:1209–1222. doi: 10.1016/S0020-7403(98)00002-2

chapter four

Design, manufacturing, and machining trials of magnesium-based hybrid parts

Eva María Rubio Alvir, José Manuel Sáenz de Pipaón, María Villeta, and José Luis Valencia

Contents

4.1 Introduction

Most parts of the aeronautical and automotive industries, in addition to satisfying the requirements of precision and quality, need to be manufactured with materials that are rigid, resistant, and lightweight. Occasionally, the available materials do not simultaneously satisfy all the required properties, so, in the design of pieces in these fields, it is necessary to adopt compromise decisions that sacrifice some of the mentioned properties.

The traditional procedure to improve the performance of materials is the development of new metal alloys or polymers that increase the areas of application, but this development can be expensive and not always achievable. An alternative is to combine two or more existing materials so as to achieve the superposition of their properties, thus giving rise to hybrid parts that can satisfy a greater range of properties.

A hybrid part means the combination of two or more materials in a certain geometry and scale, optimally for a specific engineering purpose (Ashby and Bréchet 2003; Kickelbick 2007).

In the last years, applications of hybrid parts have appeared for aluminium and steel (Wagner et al. 2001), aluminium and titanium (Luo and Acoff 2000; Wagner et al. 2001) or aluminium and magnesium (Ben-Artzy et al. 2002; Tharumarajah and Koltun 2007), and steel and magnesium (Qi and Song 2010; Nasiri et al. 2011, 2013), as well as studies to analyze the possibilities and limitations of different technologies to weld dissimilar materials (Casalino 2017; Casalino et al. 2017; Yan et al. 2017; Zakaria and Yazid 2017).

Some examples of application of hybrid parts can be found in the aeronautical sector, in the fuselage and wing building (Gururaja and HariRao 2012), where sandwich structures with sheet metal and composite materials are used (DRL 2011; Lee et al. 2013), and in the automotive field, where, in addition to metals and composite combinations (Dau et al. 2011; Frantz et al. 2011; Amancio 2012; Suryawanshi and Prajitsen 2013), other material combinations were studied, such as aluminium and titanium (Luo and Acoff 2000; Wagner et al. 2001), aluminium and steel (Wagner et al. 2001), and aluminium and magnesium (Aghion et al. 2003; Krebs et al. 2005; Staeves 2005; Fischersworring-bunk et al. 2006; Mg Showcase 1 2007).

Therefore, it is increasingly common to have to machine this new type of hybrid materials, both for their manufacturing and their repair or maintenance, and there is a new challenge in determining and optimizing the parameters involved in the cutting processes since, although there are recommendations of use for the different individual components, it may be that these are not the most appropriate when the materials are mechanized forming part of a set (Uthayakumar et al. 2008; Manikandan et al. 2012). Therefore, to optimize the machining processes of the hybrid parts, it will be necessary to carry out trials that allow studying the different variables that influence the final result of such processes and their possible interactions. A preliminary step, before making the trials, is to define the workpieces where the tests will be carried out.

There are different national and international standards for different purposes, such as the standard for steel and steel products (ISO 377 2013) for the location and preparation of samples and workpieces for mechanical tests, tests such as traction, bending by shock, crushing, flared, or curved. In the case of turning, there are the standards for steel tests (UNE 36423 1990) for the determination of the machinability index with short-term turning methods; the standard (UNE 16148 1985) for duration tests of unique cutting tools in long-duration tests at cutting speed constant; ISO 3685 (1993) for tool duration tests or ISO 243 (2014), which specifies the types and the dimensions of turning tools with carbide tips for external tools; and ISO 514 (2014) for internal tools. For milling operations, there are ISO 8688-1 (1989) for testing of tool life in planning operations on steel and casting parts and ISO 8688-2 (1989) for testing of tool life in finishing operations on steels and foundries. In the case of the machining of hybrid parts, there is no rule that defines the pattern or delimits the proportions that the workpiece must keep for testing. For this reason, given that the hybrid parts are presented as a solution for some of the problems found in the manufacture of parts in the industrial sectors that are being analyzed, it is decided to design different models of workpieces for turning, milling, and drilling that simulate being parts of hybrid parts, in order to be able to compare and determine which are the most suitable to carry out machining tests, as well as being able to analyze the different processes and look for the best cutting conditions (cutting parameters, tools, lubricants) for their machining. In general, it is usually considered that a machining process is carried out under the best cutting conditions when it is able to provide parts with the design requirements in the shortest possible time. In automotive and aeronautical industries, these design requirements usually refer to very demanding high levels of precision and quality of the pieces obtained. One of the parameters most used to measure the quality of machining is the surface roughness or microgeometric surface state of the pieces. This is related to the capacity of lubrication; friction; resistance to fatigue, corrosion, and wear; and external appearance of the pieces (Carro 1998). The decrease in the degree of surface finish of the pieces does not always improve the functioning, and in general, it is more expensive the lower it is; so it is neither useful nor economical to achieve narrower tolerances than those required in the plan (Villeta et al. 2012).

If the real surface of a piece is observed with sufficient magnification, an irregular profile can be seen that cannot be defined by a single parameter capable of giving a measure of all the geometric characteristics. However, certain parameters can be defined and measured that provide sufficient information about the characteristics that the piece will present in its later industrial operation, such as, for example, the maximum height of the profile, the average height of the irregularities, arithmetic mean roughness, mean square deviation, profile asymmetry, or profile flattening (Perez 1998). Undoubtedly, although there are other parameters to define the surface finish state, the most general is the arithmetic average roughness, Ra, expressed mathematically by means of Equation 4.1, where l is the evaluation length and $z(x)$ the ordinates values in the sampling evaluation (ISO 4287 1997; ISO 4288 1998).

$$Ra = \frac{1}{l} \int_0^l |z(x)| \, dx \qquad (4.1)$$

The degrees of arithmetic mean roughness, Ra, can be very wide depending on the type of machining, tools, and cutting conditions used. A classification broadly used is defined by the American Society of Mechanical Engineers (ASME) in the ASME B46.1 (2009) standards. Table 4.1 shows the degrees of roughness used according to the ASME B46.1 standard in the comparison of roughness samples, as well as the equivalent ISO class according to ISO 1302 (2002).

Therefore, it remains necessary to design and manufacture workpieces of hybrids parts for the performance of machining trials (Saénz de Pipaón 2013) to determine the best cutting conditions of these new combinations of materials to achieve certain quality requirements (generally superficial) specified in the drawings of the pieces. This work shows the design, manufacturing, and machining trials of magnesium-based

Table 4.1 Degrees of classification
of the arithmetic mean roughness, *Ra*

μm	μ inch	ISO Class
0.006	0.25	
0.0125	0.5	
0.025	1	N1
0.05	2	N2
0.1	4	N3
0.2	8	N4
0.4	16	N5
0.8	32	N6
1.6	63	N7
3.2	125	N8
6.3	250	N9
12.5	500	N10
25	1000	N11
50	2000	N12
100	4000	
200	8000	
400	16000	

hybrid parts, which allow analyzing their behavior in this type of processes. Additionally, an experimental study of the milling processes has been carried out as an example of the use that could be made with such as workpieces. Specifically, typical repair and maintenance operations of the aeronautical industry carried out under low speed conditions have been simulated.

4.2   Design

The ideal design of the workpieces of hybrid parts should be as close as possible to the real parts so that the results obtained in the machining tests could be easily extrapolated. However, given the great variety and complexity of shapes and sizes of parts in the industries analyzed, it is difficult to propose one or a few workpiece geometries that cover all possible alternatives.

Therefore, it is decided to design workpieces with geometries that are simple to manufacture and that allow easy attachment to the machine tool during machining tests (with standard tools) and an adequate data collection of the selected surface roughness parameters. Thus, the workpiece will have a base and one or several inserts of different materials.

4.2.1   Design considerations

4.2.1.1   Geometry

Each one of the designed workpieces will have a base and one or more inserts. The geometry of the base will depend on the machining operation to be tested and the type of machine tool where the tests will be carried out. For the operations of turning (mainly horizontal and facing turning), the base of the test piece will be a cylindrical piece, with a part that allows its attachment to the plate of claws of the machine and another where different materials will be inserted and where the trials will be carried out.

For the operations of milling (basically operations of cylindrical and frontal milling), the base of the test piece will be a parallelepiped that can be fastened with the clamp of the milling machine or machining center where the trials will be carried out and in which the inserts will be mounted.

Finally, for the drilling operations (drilling, enlarging of holes, and reaming), the parallelepiped shape is selected for the bases of the workpieces, opting also for its ease of manufacture and lashing.

The inserts of the workpieces will be made in the simplest possible geometric shapes so as to speed up and simplify their manufacture. Between these simple forms are the pieces of rectangular, square, circular, and trapezoidal section (Figure 4.1).

The pieces with rectangular, square, and trapezoidal section can be machined in milling machines; the first two offers great simplicity and ease of manufacture in this type of machines. Those with a trapezoidal section have a slightly higher difficulty in manufacturing but, in return, allow a better fixation to the base even without using specific means of joining. Finally, with respect to the pieces of circular section, they are mechanized with great simplicity in horizontal lathes and, properly positioned in the bases of the workpieces, can present improvements in front of the subjection to the base of the rectangular ones and the square ones, similar to what happens with the trapezoidal ones.

For the choice of the dimensions of the bases and inserts, in addition to the dimensions and operating characteristics of the measuring equipment, other factors will be considered, such as the dimensions of the machine tools and their means for fixing and positioning, the tools

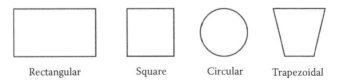

Rectangular   Square   Circular   Trapezoidal

Figure 4.1 Possible insert geometries for workpieces of the hybrid parts.

to be used, and also the raw materials provided by the suppliers, from which the bases and inserts for the manufacture of the workpieces will be obtained.

4.2.1.2   Location

The location of the inserts in the base will depend on the type of machining processes (turning, milling, or drilling), but it will be done in such a way that the tool in the advance movement mechanizes the material of the base, of the insert, and again of the base. This will allow studying the surface finish before the insert, in the insert, and after the insert.

In the turning workpieces, the inserts can be positioned radially or longitudinally. It is important that the position of the inserts allows performing the greatest number of machining operations. The location of the inserts along a generatrix of the cylindrical base is the more suitable for carrying out turning operations (horizontal and facing). This allows putting several inserts in the same workpiece. Thus, one insert can be placed, two inserts separated by 180°, three inserts separated by 120° apart, four inserts separated by 90°, or a larger number if desired (Figure 4.2).

For the milling workpieces, there are also different design alternatives in terms of the locations of the inserts in the base. Taking into account that the most common milling operations are front and cylindrical (or tangential) milling, it is important that these operations can be executed without having to make movements in the workpiece or making the minimum necessary. A possible solution is to place inserts in independent bases is shown in Figure 4.3.

Another alternative in the design of milling workpieces is to put the inserts of different materials and section types in the form of a matrix

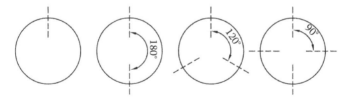

Figure 4.2 Possible insert locations for turning workpieces.

Figure 4.3 Workpieces for milling with different types of inserts in several locations.

Figure 4.4 Workpiece for milling in forming of matrix.

(rows and columns), which allows putting from 1×1 to $n \times n$ inserts in the same base (Figure 4.4).

The separation between inserts will be such that the machining of each material and section type can be carried out separately. For this, the diameter of the tool or tools with which the trials are intended will be taken into account. This will make it possible, as in the turning, to study the surface characteristics before the insert, on the insert, and after the insert.

In addition, this matrix configuration allows other more complex studies in which the surface roughness results can be compared for different types of section of the inserts and the same material or the same type of section and different materials depending on the analysis made in each moment.

For the drilling workpieces, the simplest location for the inserts will consist in alternating the different materials so that the surface roughness can be taken before, on, and after the inserts. A possible solution is the one shown in Figure 4.5a.

In this configuration, the parallelepiped inserts are interspersed with the base material. This solution, which is simple to manufacture and allows carrying out a comparative study of roughness for different materials, presents some drawbacks, such as having to use very long drill bits, performing deep drilling operations, difficulty in taking roughness

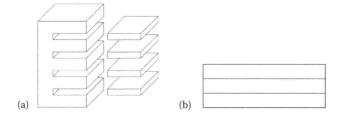

Figure 4.5 (a) Interleaving of inserts of different materials and (b) alternating base-material and insert-material in drilling workpieces.

measurements in the inserts placed in the center of the workpiece, or using fastening systems for the assembly and disassembly of the inserts that allow them to be extracted from the workpiece for making the roughness measurements.

With the intention of avoiding the aforementioned problems, a solution is shown in Figure 4.5b, in which the alternation of materials and the simplicity of design and manufacture of the base and the inserts are maintained.

In this case, all the parts that make up the workpiece have parallelepiped shape and the same dimensions, since this form is the most suitable for a correct and secure fastening to the machine. The base will be considered to be the pair of upper and lower parallelepipeds, which must be of the same material, and the insert, the one located between the previous ones and of different material. It is important that both the base and the insert can be removable so that measures can be taken in the mechanized zone with relative ease.

4.2.1.3   Materials

As mentioned earlier, the reduction of weight in the aeronautical and automotive industrial sectors is very important for both economic and environmental reasons. Therefore, lightweight materials, with a good weight/mechanical resistance ratio like light alloys and composite materials, have been introduced into the manufacture of vehicles and aircrafts, mainly in the last years. Consequently, for the design and manufacture of workpieces of hybrid parts, it will be necessary to know their properties, in particular, those related to mechanical strength, density, and machinability, as well as other aspects related to them, such as ease of acquisition or cost. Table 4.2 summarizes some properties (density, modulus of elasticity, elastic limit, elongation, and Poisson and expansion, coefficients) of steels, light alloys (most used in aeronautic and automotive industries such as

Table 4.2 Properties of some structured materials

Material	Density (kg/m³)	Modulus of elasticity (GPa)	Elastic limit (MPa)	Elongation (%)	Poisson's coefficient	CTE (10⁻⁶ m° C⁻¹)
Steels	7900	210	220–1600	10–33	0.29	11.5
Titanium	4500	120	780–1170	10–18	0.33	8.6
Aluminium	2700	71	100–500	5–19	0.33	22.9
Magnesium	1700	45	70–270	3–23	0.35	26
CFRP	1600	220	1400/38[a]	0.8/0.6		−0.2/0.3
ABS	1060	2.70	32–45	15–30	0.50	60–110

[a] Tensile strength: longitudinal direction/transversal direction.

titanium, aluminium and magnesium), as well as carbon fiber reinforced polymer (CFRP) and thermoplastic acrylonitrile butadiene styrene (ABS) (Riba 2008). The importance of light alloys can be appreciated if the relative magnitudes of their properties are considered, that is, the result of dividing such properties by density. As well, the strong anisotropy of CFRP can be seen, showing its best properties in the direction of the fibers or longitudinal direction.

Among the intrinsic characteristics of the materials are the physical properties and the mechanical properties, such as the traction resistance, compression and shear, impact resistance, ductility, tenacity, hardness, and machinability, among others.

Among them, machinability is a characteristic of each material whose knowledge will be very useful, in general, for planning production and conducting economic analysis of manufacturing processes and, in particular, for determining the best combination of materials for the manufacture of hybrid parts workpieces. The current manufacturing processes are subject to a constant improvement in productivity and cost reduction while maintaining the quality of the final product, so continuous innovation is necessary in this type of process (Qehaja et al. 2012).

In general, machinability is defined in terms of surface finish, difficulty for the chip control, strength, and power for cutting and tool life. So, the improvement in machinability can be translated to improvements in the surface finish, the chip control, the tool wear, and the cutting force reduction.

It is not easy to determine to what extent each of these factors must be considered, so in order to adopt a machinability index, it is necessary to resort to comparative methods. AISI steel SAE B1112 (desulfurized), with hardness of 160 HB and machinability of 100% (Kalpakjian and Schmid 2008), was chosen as standard.

To determine the Machinability Rating (MR) of a material, the cutting speed obtained in a test for a tool life of 60 minutes is divided between the cutting speed of the steel AISI SAE B1112 according to Equation 4.2. The value can be expressed as a percentage, as indicated by some tool manufacturers in their catalogues (Seco 2012).

$$MR = \frac{V_{60}}{V_{60(B1112)}} 100 \qquad (4.2)$$

4.2.1.4 Joints

Another factor to consider in the design of hybrid workpieces is the joint of the insert to the base. These joints can be classified into two types, fixed joints and removable joints (Table 4.3). The disassembly of the parts that

Table 4.3 Fixed and removable joints

Fixed joints	Removable joints
Rivets	Threaded elements
Welding	Prisoners
Press fit	Pins
Adhesives	Cotter pins
	Fluted shafts
	Guides

form the workpiece during its useful life is not contemplated in the fixed joints, and it is necessary to break the joining zones for their separation. On the contrary, in removable joints, it is possible to separate the pieces easily without breaking the pieces or the means that unites them.

Among the fixed joints, those that could be applied in the design of the hybrid part workpieces are press fit, welding, and adhesives:

- Press fit is the joint in which the shaft (insert) is larger than the hole (base) where it is placed. With this union, movement between both parts is prevented. To insert one piece into another, it is necessary to expand the hole by providing heat to introduce the shaft with ease. Once the base is cooled to room temperature, the insert is firmly attached to it, making the adjustment.
- Welding can be homogeneous, with the contribution of material similar to the piece being welded or without contribution, and heterogeneous, in which a material different from that of the piece to be welded is provided. The welding of the inserts to the base in the workpieces of hybrid parts can be done both longitudinally and transversely.
- Adhesives consist of interposing between the surfaces that are desired to join, a layer of material with a high adhesion power that can be natural and synthetic. The synthetic ones are able to withstand large loads and have good resistance to cutting forces and fatigue, as well as environmental exposure (corrosion, high temperatures, and chemical attack). Three main types exist: epoxies, urethanes, and acrylics.
  – Epoxies have a high adhesion, present the strength at high temperature, and combine metals, plastics, and rubber. Their main disadvantage is they require more time than the other types since surfaces must be prepared and the adhesives have to be cured.
  – Urethanes have good flexibility and impact resistance; are suitable for joining plastics, rubber, and wood; and have a lower final adhesion than epoxies, and their use in bare metals and crystals is not recommended.

– Acrylic adhesives have very good resistance to temperature, shear stress, and peeling in both metals and plastics; present an excellent final adhesion by quickly joining most metals; and are easy to apply, and their curing time is low. Their main disadvantage is that when they cure, they contract up to 7% (Henkel 2011; Ruiz 2011).

The demountable joints that have good application in the hybrid parts workpieces are threaded elements. Among the different threaded elements are, for example, screws, nuts, studs, and inserters. These threaded elements are the fastening system most used in the union of pieces that can later be disassembled. The inserts can be attached to the base by means of screws or inserters threaded according to the final configuration of the workpiece.

4.2.1.5   *Other considerations*

Other considerations to be taken into account in the manufacture of the workpieces of hybrid parts are the ease, or not, of obtaining raw materials; the available resources; the additional costs due to, for example, the means of joining (welding, threaded elements, or adhesives) and if they require specific material resources (furnaces, machines, filler materials, etc.) or operations to prepare the surfaces; and the possibility that galvanic corrosion appearing in some combinations of materials with very different electrochemical behaviors will reduce the life of the workpieces. This can be minimized by avoiding the combination of highly reactive materials or, if this is not possible, using a thin layer of insulation between the inserts and the base. The bonding by means of adhesives can provide such insulating layer at the same time as it serves for the union of the insert to the base and the possible loss of the joining means (threads, welds, and adhesives) during the machining trials could endanger the integrity of the workpieces. This can be avoided by properly defining and dimensioning the workpieces.

4.2.2   *Turning workpieces*

Described here are the different types of workpieces that have been designed for the turning trials. Different cases have been raised on a common cylindrical basis. Specifically, three different types of inserts have been chosen, with a section in rectangular, circular, and trapezoidal shape being denominated, respectively, type 1, type 2, and type 3.

From the different location proposals for the inserts in the base, described in the *Location* section, it is considered that the most suitable is that of two inserts of the same material 180° apart, since this number of inserts and their arrangement allow making measurements of the surface roughness before the insert, in the insert, and after the insert in different positions and thus have a comparison between them (which would not be possible with a single insert). In addition, the manufacture of the workpiece with two inserts is easier and cheaper than if it were made with a greater number of them. The fact is that introducing two inserts of the same material will allow knowing better the joint behavior of each pair of materials than if they were introduced more than two simultaneously.

Magnesium alloy UNS M11917 is taken for the manufacture of the base, and for the inserts, aluminium alloys UNS A92024, titanium alloy UNS R56400, and the steel UNS G11170.

Magnesium is taken as material for the manufacture of the bases due to its lightness and ease to be machined, which will result in lightweight, easy-to-handle workpieces that will require less power (energy) to manufacture.

Table 4.4 shows the chemical composition of the chosen materials and Table 4.5 shows some of their properties. With the intention of unifying the denomination of the materials, the United Number System (UNS) is collected along with its denomination in other well-known systems.

Inserts will be joined to the base by means of an adhesive, preventing, like this, the possible galvanic corrosion problems that usually appear when joining metals with different electrochemical behaviors (in particular in the combination of magnesium and titanium) is prevented. In addition, joins will be easy and simple to make, and they will provide a longer life of the workpiece, that is, the possibility of making a greater number of trials in safety conditions.

The common base of the workpieces is a cylindrical piece where the recommendations given in the standard UNE 16148 (1985) for the

Table 4.4 Chemical composition (%) of the materials used in the manufacture of the workpieces

UNS M11917 (AZ91D)	UNS A92024 (AA2024 T351)	UNS G11170 (AISI/SAE 1117)	UNS R56400 (Ti-6Al-4V)
Al 8.30–9.70	Al 90.7–94.7		Al 5.5–6.75
Cu ≤0.03	Cr ≤0.1		C ≤0.08
Fe ≤0.005	Cu 3.8–4.9	C 0.14–0.2	H ≤0.015
Mg 90	Fe ≤0.5	Fe 98.33–98.78	Fe ≤0.4
Mn ≥0.13	Mg 1.2–1.8	Mn 1–1.3	N ≤0.03
Ni ≤0.002	Mn 0.3–0.9	P ≤0.04	O ≤0.2
Si ≤0.1	Si ≤0.5	S 0.08–0.13	Ti 87.725–91
Zn 0.35–1	Ti ≤0.15		Zn 3.5–4.5
	Zn ≤0.25		

Table 4.5 Properties of the materials used in the manufacture
of the workpieces

	UNS M11917 (AZ91D)	UNS A92024 (AA2024 T351)	UNS G11170 (AISI/SAE 1117)	UNS R56400 (Ti-6Al-4V)
Density (kg/m³)	1810	2780	7850	4430
Brinell hardness (HB)	63	120	121	334
Resistance to tension (MPa)	230	469	425	900
Elastic limit (MPa)	150	324	285	830
Elongation (%)	3	19	33	10
Modulus of elasticity (GPa)	44.8	73.1	205	114
Poisson's coefficient (ν)	0.35	0.33	0.29	0.33
Shear modulus (GPa)	17	28	80	44
Thermal conductivity (W/mK)	72.7	121	49.8	6.7
Specific heat (J/g°C)	1.047	0.875	0.481	0.5263
Thermal expansion Coefficient ($\times 10^{-6}$m°C^{-1})	26	22.9	11.5	8.6
Melting point (°C)	595	638	1430	1660

length–diameter relation have been followed. It consists of two different diameters, one of 53 mm and length of 60 mm and another of 32 mm and length of 18 mm (Figure 4.6).

These dimensions are determined taking into account the dimensions of the starting material, the characteristics of the machines and equipment where the workpieces will be manufactured, the machining trials number to perform, and the way to do measurements of the surface roughness. A total of nine bases will be made, three for each type of workpiece (type 1, type 2, and type 3).

4.2.2.1 Type 1 workpiece

From the common base described earlier, 2 holes for the inserts of the different types of workpiece will be made. They will be separated from each other by 180°. Type 1 will have a rectangular section of dimensions 10 × 10.5 × 60 mm, as can be seen in Figure 4.7.

Among the advantages presented by the type 1 workpiece is the low manufacturing difficulty of the holes in the common base and of the inserts. Another advantage of the type 1 workpiece is that the inserts cannot be turned by the action of the cutting forces during the turning trials that are carried out. Among the disadvantages is the possibility that

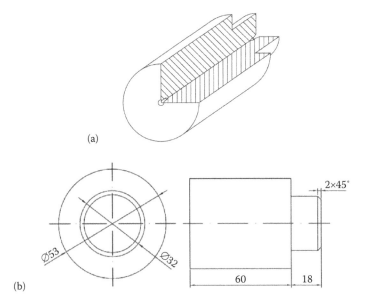

(a)

(b)

Figure 4.6 Common base of the turning workpieces. (a) Conceptual idea. (b) Design plane.

the inserts come out due to the centrifugal force generated in the rotation speed of the workpiece.

4.2.2.2 Type 2 workpiece

In the type 2 workpiece, the holes for the inserts will be 15 mm in diameter, circular section, and 60 mm in length, and it will be tangent to a generatrix of the common base and separated by 180° from each other (Figure 4.8).

The advantage of the type 2 workpiece is that it is more difficult for the inserts to come loose during machining due to the centrifugal force generated with the spindle speed of rotation. The drawbacks of this type 2 workpiece are the greater difficulty in machining the holes for the inserts, the cutting forces generated during machining can cause the inserts to turn, and the impossibility of using the entire insert to carry out tests, since from the moment that two diametrically opposed points are machined, the degree of fixation to the base decreases and, therefore, the safety of the machining decreases.

4.2.2.3 Type 3 workpiece

In the type 3 workpiece, the holes for the inserts will be of a trapezoidal section of dimensions 20 × 8.5 × 60 mm with two 60° angles, and as in the previous cases, they will be 180° apart (Figure 4.9).

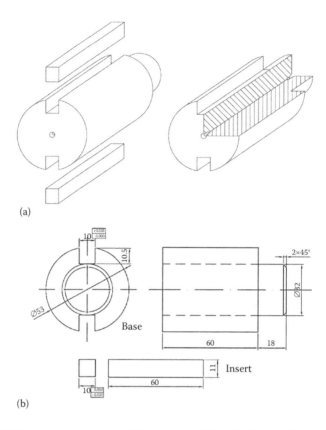

Figure 4.7 Workpiece of type 1. (a) Conceptual idea. (b) Design plane.

The advantage of this type 3 workpiece is the combination of the disadvantages of the type 1 and 2 workpieces discussed previously. During the machining trials, the inserts will not be able to come off due to the centrifugal force resulting from the rotation speed, nor can they be rotated inserts as a consequence of the cutting forces generated during machining. The disadvantage of this type 3 workpiece is that the execution times will be longer than in the other types proposed, since the geometry of the inserts is more complicated than in the inserts with rectangular and circular sections.

4.2.3 Milling workpieces

Following the methodology used in the turning workpieces and the proposals described in the *Geometry* section, three types of geometries (rectangular, circular, and trapezoidal sections) will be used for the inserts in the milling workpieces. The inserts of rectangular and trapezoidal sections allow carrying out both operations of face milling and cylindrical

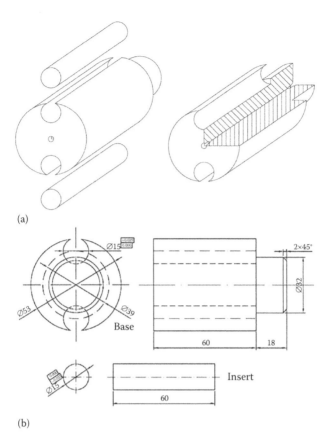

(a)

(b)

Figure 4.8 Workpiece of type 2. (a) Conceptual idea. (b) Design plane: base and insert.

or tangential milling without the need to move the workpiece; that is, it is not necessary to untie the workpiece and hold it in a different way, while inserts with a circular section are more suitable for face milling operations but also allow cylindrical milling.

The materials used in the workpieces for milling, and the reasons for their selection, are the same as those of the workpieces for turning. That is, for the manufacture of the base, magnesium alloy UNS M11917 will be used, and for the inserts, aluminium alloy UNS A92024, steel UNS G11170, and titanium alloy UNS R56400 are used. The locations of the inserts to the base will be done following the approach in a matrix form described in the *Location* section, concretely, in the form of a 3 × 3 matrix that will allow the placement of the three types of inserts (Figure 4.10).

For the determination of the dimensions of the base and of the holes for the inserts, the same factors as for the turning workpieces have been

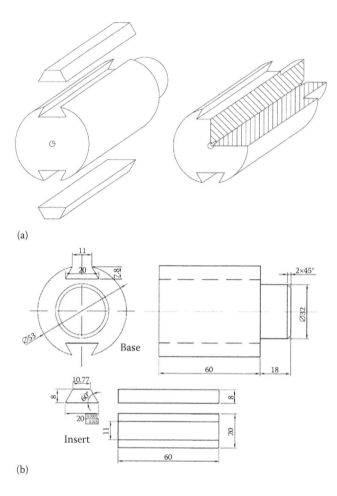

Figure 4.9 Workpiece of type 3. (a) Conceptual idea. (b) Design plane: base and insert.

taken into account. That is, the dimensions of the raw material and the characteristics of the machines and equipment where workpieces will be manufactured, machining trials, and surface roughness measurements are to be carried out.

The base will be defined by a parallelepiped of dimensions 110 × 70 × 50 mm on which the holes will be machined for the inserts of a rectangular section (locations {1,1}, {1,2}, {1,3}), a circular section (locations {2,1}, {2,2}, {2,3}), and finally, a trapezoidal section (locations {3,1}, {3,2}, {3,3}). The dimensions for the different geometries of the holes will be the same as the inserts.

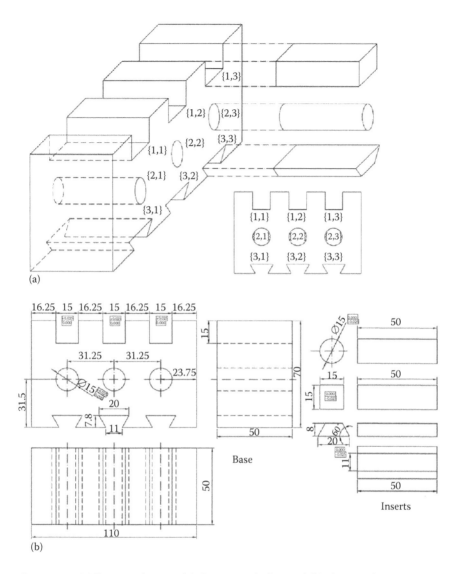

Figure 4.10 Milling workpiece. (a) Conceptual idea and (b) design planes.

That is, the dimensions of the rectangular section inserts will be 15 × 15 × 50 mm, those of the trapezoidal section inserts will be 20 × 7.8 × 50 mm, and finally, those of the circular section inserts will be 15 mm in diameter, and all them are 50 mm in length.

The inserts joining to the base will be made by using an adhesive as in the turning workpieces, since it is easy and simple while it can reduce the galvanic corrosion problems.

4.2.4   Drilling workpieces

The drilling workpieces for hybrid parts will consist of intercalating, in a base material, different materials so that the drill passes from one material to another during machining. However, as discussed in the *Location* section, the interleaving of more than one material in a base, although it is simple to manufacture and allows the comparative study of the roughness for different materials, may not be suitable because it involves the use of long drill bits, demanding deep drilling operations or requiring the measurement of roughness in difficult-to-reach points with the available measuring equipment. For this reason, it has been decided to define the drilling workpieces of hybrid parts as shown in Figure 4.11, where the base will be formed by the upper and lower parallelepipeds and the insert will be placed between them. The materials used in the manufacture, and the reasons for their selection, are the same as for the turning and milling workpieces. Therefore, the bases of the drilling workpieces will be two parallelepipeds of magnesium alloy UNS M11917 of dimensions of 50 × 50 × 15 mm and, as inserts, parallelepipeds of aluminium alloy UNS A92024, steel UNS G11170, and titanium alloy UNS R56400 of the same dimensions as the base, 50 × 50 × 15 mm. A total of three workpieces will be manufactured (one for each insert). The fixation of the three parallelepipeds that make up each of them will be mechanical, so that the surface roughness inside the machined holes can be disassembled and

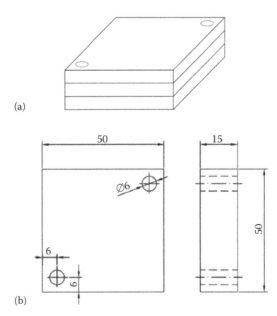

Figure 4.11 Drilling workpiece. (a) Conceptual idea and (b) design planes.

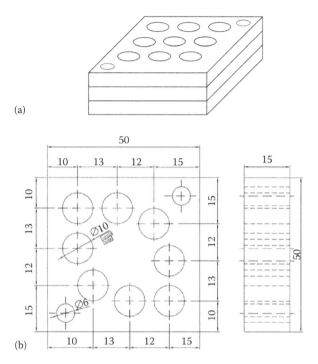

(a)

(b)

Figure 4.12 Drilling workpiece for the experimental study of the repair and maintenance of holes. (a) Conceptual idea and (b) design planes.

measured with relative ease. In order to simulate repair and maintenance operations, a workpiece will be predrilled and will have a number of holes in accordance with the experimental test planned for it.

In certain cases, it is not important to perform a complete drilling, as in the case of hole repair and maintenance operations, where small depths are used, so that the workpieces can be predrilled in order to simulate such operations (Figure 4.12).

4.3   Manufacturing

4.3.1   Previous activities to manufacturing

To make the manufacturing as efficient as possible, both the machining of the individual elements and the assembly and bonding, it is necessary to carry out a series of previous activities, such as the collection of materials, tools, and necessary fixtures, as well as the verification of the availability and state of operation of the machines and equipment where the manufacture of the workpieces will be carried out.

In addition, before the execution of each operation, it will be necessary to have, as close as possible to the machine, all the elements necessary for assembly and mechanized and dimensional control of the manufactured parts, since, in parallel to the manufacture of the workpieces, a study is going to be carried out on times and difficulty of manufacture that allows determining which are the simplest and fastest workpieces to manufacture. It is true that both the time of manufacture and their difficulty will depend, to a large extent, on the supply dimensions of raw materials, for which they must be acquired with the measures closest to those of use to reduce the machining time.

In this case, the raw materials were acquired in the following dimensions: UNS M11917 magnesium alloy in ingots form 55 × 115 × 700 mm, UNS A92024 aluminum alloy in bars of 1 m of length and circular section of 25 mm and 75 mm diameter, respectively, UNS R56400 titanium alloy in bars of 1 m of length with a circular section of 75 mm diameter, and UNS G11170 steel in a bar of 1 m of length with a rectangular section of dimensions 50 × 30 mm.

A list of the tools and the measuring instruments was made with adequate dimensions and scales for the workpiece manufacture, namely, toolholders, tool inserts, milling tools, drill chuck, center drill bit, drills bit, reamers, and fine file and sand paper for removing sharp edges. As well, for the dimensional control, a Vernier caliper, a micrometer, and a dial indicator are needed. All of these, along with the raw materials, were placed near the workplace, at the necessary moment for manufacturing. Once the different components of the workpieces were manufactured, they were assembled and/or glued. The next paragraphs show how the workpieces were manufactured.

4.3.2   Manufacturing of turning workpieces

This section will show how to prepare the common base of UNS M11917 magnesium alloy workpieces and the inserts (in UNS A92024 aluminium alloy, UNS R56400 titanium alloy, and UNS G11170 steel) for each of the three types of workpieces manufactured type 1 (with inserts of rectangular section), type 2 (with inserts of circular section) and type 3 (with inserts of trapezoidal section). In addition, the joining means used on each of the workpieces and a comparative analysis thereof are shown.

4.3.2.1   Machining of the common base

The total number of workpieces to be manufactured is nine, so it is necessary to make nine UNS M11917 magnesium alloy bases. The machining operations in this first process are carried out entirely in the horizontal lathe (Figure 4.13).

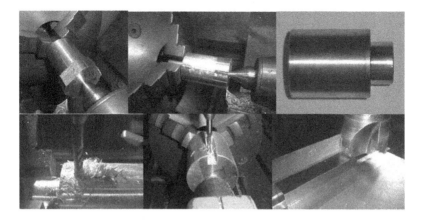

Figure 4.13 Machining workpiece type 1.

4.3.2.2   Type 1 workpiece

The manufacture of this type of workpiece is done in the universal milling machine. For the machining of the common base of magnesium alloy, a dividing device, tailstock, a chuck holder, an 8-mm diameter straight hard milling cutter, a dial indicator, and centering are necessary (Figure 4.13).

For the machining of the inserts, a clamp and a plate of interchangeable plates with a diameter of 80 mm are necessary. A fine file is also necessary to eliminate burrs generated during machining. The dimensional control is carried out by a Vernier calliper and micrometer.

4.3.2.3   Type 2 workpiece

The bases of type 2 workpieces are mechanized in the universal milling machine while the inserts are in the horizontal lathe. For machining in the milling machine, claw plate, drill chuck, drill bits (with diameters 6 mm and 13 mm), and tapping are necessary, in addition to a probe, a dial indicator, and centering (Figure 4.14).

Due to the fact that a drill with a length enough is not available for the machining of circular section holes, a cobalt tool is prepared that enables machining with the help of a mandrel. Another problem that arises is the exit of the tool, since the claws of the plate prevent it. To solve this problem, an aluminium clamp is manufactured on the horizontal lathe (Figure 4.14), which allows the base to be clamped and the tool to be exited. The necessary measuring instruments are a 0–25-mm micrometer, 12–25-mm comparator, and Vernier calliper, all with centesimal accuracy.

The machining of the inserts is done in the horizontal lathe and the necessary tools are tool holders with interchangeable insert, cutting tool, chuck, and drill bit. The measuring devices used during

Figure 4.14 Machining the workpiece type 2.

machining are Vernier calliper and 0–25-mm micrometer. As in the type 1 workpiece, a fine file is necessary to eliminate the burrs generated during machining.

4.3.2.4   *Type 3 workpiece*

The manufacture of both the base and the inserts of the type 3 workpiece is done in a universal milling machine (Figure 4.15). Among the necessary tools for machining are dividing device, counterpoint, clamp, toolholder, interchangeable plates 80 mm in diameter, straight (8 mm and 10 mm) and angular (60°) carbide mills, dial indicator, centering, and a fine file. The measuring devices required for the control of the dimensions are a Vernier calliper and 0–25-mm micrometer.

For the machining of trapezoidal section inserts, parallel guides are used where the starting material is located for its machining and is tilted to the machine at a 60° angle. A 10-mm-diameter mill is held in a tool holder and the machining is performed.

Figure 4.15 Machining the workpiece type 3.

4.3.2.5   Assembly and gluing of the workpieces

Although the design of the different elements has been made so that the union of the inserts with their holes was adjusted, for greater security during the performance of the trials, a choice was made to use joining for the fastening of the inserts to the base. These consist of a fixed joint that prevents the inserts from being pulled out or turned by the effect of the cutting forces and the centrifugal force resulting from the cutting conditions.

The fixed joint chosen is a specific adhesive for metals. Due to the possible heat generated in machining operations, this adhesive must have some resistance to temperature. The adhesive is a high-viscosity mono-component green liquid with anaerobic curing that also allows secondary curing by means of an activator. The commercial name is Loctite® 640 and is manufactured by Henkel Corporation. Once cured, its resistance is high. The speed of curing depends on the substrate that is used and the curing time (ISO 10123 1990). It is a product designed for the union of components between adjusted metal surfaces, avoiding loosening due to impacts or vibrations. Curing is carried out in the absence of air. For this application, this glue requires individual protection equipment, such as latex gloves or similar, in addition to goggles with side protection, as it can cause irritation in the eyes and skin, as well as allergic skin reactions.

Before bonding, the different pieces (bases and inserts) were cleaned with alcohol, using the glue for industrial use (Loctite® 640) in the gluing of the inserts to the different bases (Figure 4.16). To accelerate the bonding process, the Loctite® 7471 curing activator was used.

4.3.2.6   Analysis and compared discussion

Once the three different bases and the different inserts are manufactured, the analysis is carried out according to criteria of difficulty in the manufacture and the union as well as in the time used in the manufacture of all. Base type 1 (rectangular housing) is the one that presents the least difficulty for machining and the one that takes place in the shortest time. The inserts of this type 1 workpiece also have a low degree of difficulty. The operations of mechanization of the inserts are simple, but it is necessary to be careful so that the faces are parallel two by two at the same time that they are perpendicular to each other, since it facilitates the adjustment in the housing of the base. Table 4.6 shows the result of the times used in the manufacture of the type 1 workpiece.

With the type 2 base (circular housing), special care has been taken in centering and aligning it for machining it, since the perpendicularity of the generatrix with the plane defined by the axes x and y (milling table) is important. Although the material of the base (magnesium alloy) is not hard and is well machined, the drilling must be done progressively; if not,

(a)

(b)

Figure 4.16 (a) Gluing of the inserts to the different bases. (b) Workpieces type 1, 2, and 3 once the gluing process is finished.

when working on the edge of the base, the drill tends to move outward, since that offers less resistance, leaving a housing inclined with respect to the generatrix that may not be corrected with the machining. Another problem that occurs in the machining is the tool relieve, since the claws of the plate avoid it. To solve this problem, an aluminum clamp has been machined; it has also been necessary to sharpen a cobalt tool that allows machining of the housings. The machining of the inserts for this base is the easiest to perform and does not present difficulty for adjustment into the base. The times used in the manufacture of type 2 workpiece are shown in Table 4.7.

In the type 3 workpiece, the execution time is longer than in type 1 since once the slot has been machined in the dividing device, this has to be disassembled and a clamp is assembled for the final machining using the angular milling bit. Slot alignment is important for the correct machining of the trapezoidal section housing. The time invested in the realization of the type 3 insert is greater than in type 1, since two planes inclined at 60° must be made; however, the manufacture of this insert presents the difficulty of adjustment in the housing, being the adjustment that offers the

Table 4.6 Time analysis of workpiece type 1 for turning

Workpiece type 1 (Rectangular section)		Operations	No. of units	Time (hours)
Base	Tools:	Base turning	3	0.16
	Dividing device, tailstock, collet chuck, 8-mm straight bit, dial indicator, centering.	Assembly of the dividing device (alignment of the divider and part centering)	1	0.25
	Material:			
	Magnesium alloy	Machining	3	0.50
Total time for the manufacture of all bases				2.25
Inserts	Tools:	Assemble and align the clamp	1	0.25
	Clamp, plate of plates of 80 mm	Inserts machining		
	Material:	Aluminium alloy	2	0.50
	Aluminium alloy	Titanium alloy	2	1
	Titanium alloy	Steel	2	0.67
	Steel			
Total time for the manufacture of all inserts				4.58
Total time used in the gluing of all inserts				0.16
Total time for the manufacture of all workpieces				7

greatest difficulty of realization among the three types proposed. Table 4.8 shows the times used in the different operations carried out for manufacturing type 3 workpiece.

Finally, Figure 4.17 shows the total time used in the manufacture according to the workpiece type and the material used.

4.3.3   *Manufacturing of milling workpieces*

The manufacturing of milling workpieces is done in the universal milling machine and in the horizontal lathe. For the machining of UNS M11917 magnesium alloy base, the following are necessary: clamp, chuck, straight carbide bit of 10 mm diameter and angle of 60°, 80-mm-diameter interchangeable plates, boring tool, cobalt tool, chuck, dot drill, drills of 12 and 14.50 mm diameter, in addition to a comparator clock with oscillating probing (tracer), a centering, a rule with a 90° angle, and some parallel guides (Figure 4.18).

Table 4.7 Time analysis of workpiece type 2 for turning

Workpiece type 2 (Circular section)		Operations	No. of units	Time (hours)
Base	Tools:	Base turning	3	0.16
	Chuck, collet chuck, drill chuck, 6- and 13-mm diameter drill bits, manual tools, dial indicator, centering device, and aluminum clamp. Material: Magnesium alloy	Chuck assembly	1	0.17
		Machining	3	1.25
Total time for the manufacture of all bases				4.42
Inserts	Tools:	Lathe preparation	1	0.16
	Tool holder, parting insert, insert, drill Chuck, and center drill bit Material: Aluminium alloy Titanium alloy Steel	Inserts machining		
		Aluminium alloy	2	0.25
		Titanium alloy	2	0.67
		Steel	2	0.50
Total time for the manufacture of all inserts				3
Total time used in the gluing of all inserts				0.16
Total time for the manufacture of all workpieces				7.58

The machining of rectangular and trapezoidal section inserts is made in the universal milling machine and it requires a clamp, interchangeable plates of 80 mm diameter, and a hard metal straight mill of 10 mm diameter. The circular section inserts are made in the horizontal lathe, requiring a tool holder with interchangeable plates, cutting tools, drill chuck, and drill bit (Figure 4.19). As in the previous case of workpieces, for turning, in the manufacture of both the base and the inserts, a fine file and fine-grained sandpaper are used to eliminate the burrs generated during machining. The measuring instruments used are 150 mm callipers, 0–25 mm micrometer, and 12–25 mm comparator, all with an accuracy of 0.01 mm.

4.3.3.1 *Assembly and gluing of the workpieces*
The means used for the join as well as the procedure following are the same as those described for the workpieces for turning. Figure 4.20

Table 4.8 Time analysis of workpiece type 3 for turning

Workpiece type 3 (Trapezoidal section)		Operations	No. of units	Time (hours)
Base	Tools:	Base turning	3	0.16
	Dividing device, tailstock, collet chuck, milling vice, 8-mm straight bit and angular 60°, dial indicator, centering	Assembly of the dividing device (alignment of the divider and part centering)	1	0.25
	Material: Magnesium alloy	Disassembly of the dividing device and assembly of the milling vice (aligning and centering)	1	0.33
		Machining	3	0.42
Total time for the manufacture of all bases				2.33
Inserts	Tools:	Assemble and align the clamp	1	0.25
	Clamp, collet chuck, plate of plates of 80 mm, 10-mm straight bit	Preparation for 60° machining	1	0.16
	Material Aluminium alloy Titanium alloy Steel	Inserts machining		
		Aluminium	2	0.67
		Steel	2	0.83
		Titanium	2	1.25
Total time for the manufacture of all inserts				5.92
Total time used in the gluing of all inserts				0.16
Total time for the manufacture of all workpieces				8.42

shows the final appearance of the workpieces once the gluing process is finished.

4.3.3.2 *Analysis and compared discussion*

Once the workpiece for milling is finished, the analysis of the manufacturing difficulty and the times employed is carried out (Table 4.9).

In general, the workpieces for milling do not present great difficulty for their machining. The circular housing is the one that is easier to manufacture, since machining in the center of the base does not present problems, as what happens in the type 2 workpiece for turning. The insert with a circular section is also the easiest to machine because it involves

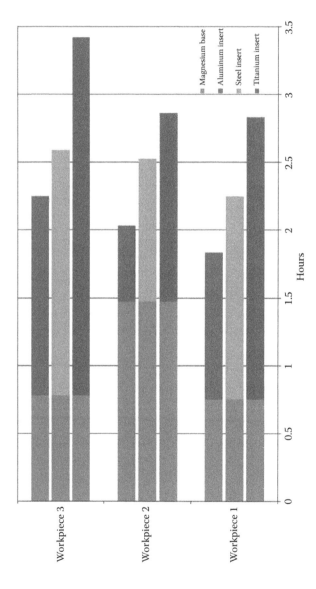

Figure 4.17 Time used for manufacturing according to workpiece and material.

Figure 4.18 Machining of the milling workpiece base.

Figure 4.19 Insert machining of rectangular, trapezoidal, and circular section.

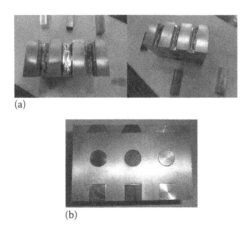

(a)

(b)

Figure 4.20 (a) Gluing of the inserts to the base. (b) Workpiece once the gluing process is finished.

Table 4.9 Time analysis of the workpiece for milling

Workpiece for milling		Operations	No. of units	Time (hours)
Base	Tools: Clamp, plate of plates of 80 mm, collet chuck, 10-mm straight bit, 60° angular bit, manual tool, mandrel, drill chuck, 12- and 14.5-mm-diameter drill bits, center drill bit, dial indicator, centering device, square, parallels Material: Magnesium alloy	Assemble and align the clamp	1	0.25
		Machining	1	0.75
		Rectangular housing machining	3	0.25
		Circular housing machining	3	0.12
		Trapezoidal housing machining	3	0.20
Total time for the base manufacture				2.70
Inserts	Tools: Clamp, plate of plates of 80 mm, collet chuck, 10-mm straight bit, tool holder, parting insert, insert, drill chuck, center drill bit. Material: Aluminium alloy Titanium alloy Steel	Assemble and align the clamp and lathe preparation	1	0.42
		Rectangular insert machining		
		Aluminium	1	0.50
		Steel	1	0.67
		Titanium	1	1
		Circular insert machining		
		Aluminium	1	0.25
		Steel	1	0.50
		Titanium	1	0.67
		Trapezoidal insert machining		
		Aluminium	1	0.67
		Steel	1	0.83
		Titanium	1	1.25
Total time for the manufacture of all inserts				6.75
Total time used in the gluing of all inserts				0.16
Total time for the manufacture of all workpieces				9.62

simple turning operations. The adjustment of the circular section insert into the base is also very simple. The rectangular section housing does not present difficulties for its machining but is where more time is invested for its manufacture (roughing and finishing). The insert with a rectangular section presents simple machining operations; however, care must be taken so that the faces are parallel two by two and perpendicular to each other to facilitate adjustment with the base.

The machining of the trapezoidal housing is easier than in the type 3 turning workpiece, since the base does not have to be moved at any time. The time invested is less than in the rectangular housing since the finishing is done in a single pass, leaving the shape of the tool. On the contrary, the trapezoidal section insert machining is in which more time is used, since the two faces inclined at 60° have to be machined. Depicted in Figure 4.21 is the time invested during manufacture according to the section shape and type of material.

4.3.4   Manufacturing of drilling workpieces

The manufacturing of the bases and inserts (Figure 4.22) for the drilling workpieces is carried out in the universal milling machine with the help of a clamp, a drill chuck, a straight 10-mm-diameter carbide mill, an 80-mm-diameter interchangeable plate, drill bits 6, 8, and 10 mm in diameter, a reamer, a handle, a dial gauge with oscillating probe (feeler), centering, a rule with a 90° angle, a protractor, parallels, and flanges. For the edge cutting and deburring, a fine lime and fine-grained sandpaper are used. In addition, to take measurements, a 150-mm Vernier calliper with an accuracy of 0.01 mm is used.

4.3.4.1   Assembly of the workpieces

According to the design planes, the drilling workpieces are assembled by removable joints, specifically, with 6-mm diameter screws and threads, as indicated in Figure 4.23.

4.3.4.2   Analysis and compared discussion

Because they are similar pieces (parallelepipeds), machining does not present great difficulty. Care must be taken that the parallelepiped faces are parallel two by two and the same time perpendicular to each other, in addition to the correct position of the hole that makes the clamping possible. Table 4.10 shows the times used during manufacture. Figure 4.24 shows the time used in manufacturing according to operation type and material used.

4.4   Trials

4.4.1   Introduction

The design and manufacturing of magnesium-based workpieces have been carried out with the aim of making machining trials. Through these trials, it is intended to deepen the knowledge on the behavior of the different combinations of materials made by machining them to different

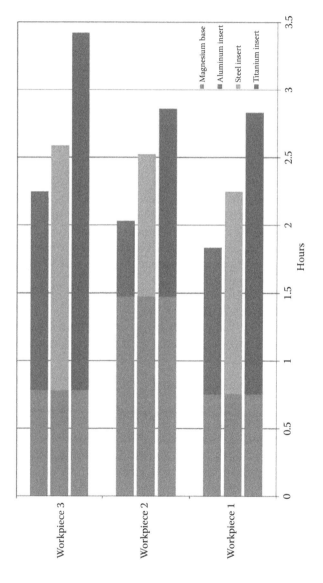

Figure 4.21 Time used for manufacturing according to shape and material.

Figure 4.22 Machining of the drilling workpieces: (a) bases and (b) inserts.

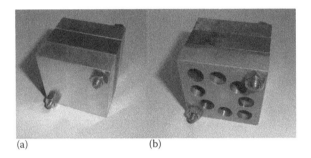

Figure 4.23 Assembly of drilling workpieces with removable joints for (a) drilling trials and (b) trials of repair and maintenance of holes.

cutting conditions, in each of the machining processes described earlier. In this chapter, attention is going to be focused on milling trials. Milling trials have been planned throughout an experimental design (Montgomery 2012). The aim was to discover the most influential factors on the surface quality of the milled workpieces, obtaining a statistical model for the prediction of the surface roughness from the values of such factors.

The following sections describe how the experimental design was planned, the material resources were used, the statistical analysis carried out, and the discussion of the results obtained.

4.4.2 *Previous activities to machining*

The milling trials are focused on repair and maintenance operations in pieces of the aeronautical sector. So, it is intended that the pieces' dimensions vary as little as possible and that the surface roughness *Ra* meets the

Table 4.10 Time analysis of the workpiece for drilling

Workpieces for drilling		Operations	No. of units	Time (hours)
Base	Tools:	Assemble and align the clamp	1	0.25
	Clamp, plate of plates of 80 mm, collet chuck, 10-mm straight bit, drill chuck, 6-mm-diameter drill	Machining	1	0.33
	bit, center drill bit, dial indicator, centering device, square, parallels	Face milling and cutting	3	0.50
	Material: Magnesium alloy	Drilling	6	0.08
Time for the manufacturing all bases				2.58
Inserts	Tools:	Assemble and align the clamp	1	0.25
	Same tools used for the base Material:	Milling of the inserts		
	Aluminium alloy	Aluminium	1	0.50
	Titanium alloy	Steel	1	0.67
	Steel	Titanium	1	1
		Drilling of the inserts		
		Aluminium	1	0.17
		Steel	1	0.25
		Titanium	1	0.42
Total time for the manufacture of all inserts				3.25
Total time used in the assembling the inserts to the base				0.08
Total time for the manufacture of all workpieces				5.92

values usually specified for this industrial sector, that is, 0.8 μm < *Ra* < 1.6 μm. This implies that the cutting conditions of the tests were carried out at levels of "low speed machining," that is, low values of the spindle speed (S), of the feed rate (f), and of the depth of cut (d). In order not to increase excessively the number of trials in the experimental design, two values of the spindle speed ($S1$ = 1910 rpm and $S2$ = 3820 rpm) and two values of the feed rate ($f1$= 250 mm/rev and $f2$ = 750 mm/rev) were selected, whereas that the depth of cut was fixed at a constant value (d = 0.25 mm) in order to maintain the dimensional requirements of the piece.

The type of tool (T) was also considered as a potential influential factor. Two types of tools of the same geometry and dimensions but different coating were selected ($T1$ = HX and $T2$ = F40M), attempting to cover the largest possible number of materials that can be machined with them.

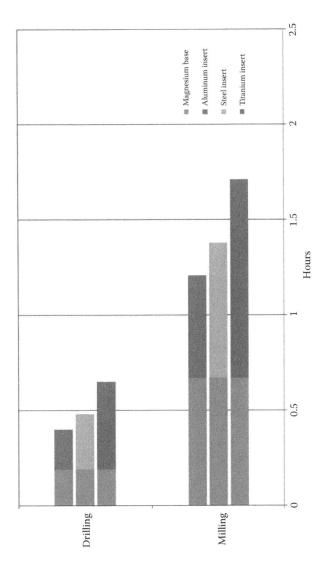

Figure 4.24 Time used for manufacturing according to operation and material.

The type of tool designed by HX is suitable for the milling of foundries, hardened steels, and nonferrous alloys, while the type of tool designed by F40M is recommended for steels, stainless steels, and heat-resistant alloys. Figure 4.25 shows an image of both type of tools, as well as their qualities.

On the other hand, due to the fact that surface roughness may vary along the workpiece (Rubio et al. 2005), it was decided to take surface roughness Ra measurements on different locations with respect to the workpiece (LRS) and on different locations with respect to the insert (LRI). In concrete, two locations with respect to the workpiece ($LRS1$ = at the beginning of the workpiece and $LRS2$ = at the end of the workpiece) and three locations with respect to the insert ($LRI1$ = before the insert, $LRI2$ = on the insert, and $LRI3$ = after the insert) were taken into account.

Factors to investigate in the milling trials and their fixed levels are shown in Table 4.11. In order to select the levels of the factors spindle speed and feed rate, in addition to attempting to simulate the operations of repair and maintenance, the characteristics of the machining center in which the trials are going to be performed (Tongtai TMV510), as well as the characteristics of the numerical control with which it is equipped (Fanuc series oi-MC), were taken into account. The machining center where the trials were made is shown in Figure 4.26a. With respect to the material

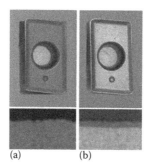

(a)   (b)

Figure 4.25 Types of coating: (a) HX and (b) CP200.

Table 4.11 Factors and levels

Factors	Levels (Notation)	Levels (Values)	
Feed rate, f (mm/rev)	$f1, f2$	250	750
Spindle speed, S (rpm)	$S1, S2$	1910	3820
Type of tool, T	$T1, T2$	HX	F40M
Location respect of the workpiece, LRS	$LRS1, LRS2$	Beginning workpiece/ end workpiece	
Location respect of the insert, LRI	$LRI1, LRI2, LRI3$	Before insert/on insert/after insert	

(a)

(b)

(c)

Figure 4.26 (a) Machining center Tongtai TMV510 equipped with CNC Fanuc series oi-MC. (b) Roughness tester Mitutoyo Surftest SJ 401. (c) Detail of the measurement process.

resources for the roughness measurements, the Mitutoyo Surftest SJ 401 was used (Figure 4.26b and 4.26c).

Taking into account the factors and its levels selected for the milling trials, an experimental design product of a full factorial 2^3 and a block of two factors (3×2) was planned. Table 4.12 shows such experimental design.

4.4.3 Milling trials

The milling trials were carried out on the workpiece for milling manufactured as indicated in Section 4.3.3, machining it according to the longitudinal direction of the rectangular steel insert UNS G11170 located on the basis

Table 4.12 Experimental design product of a full factorial 2^3
and a block of two factors 3×2

Obs.	T	S (rpm)	f (mm/rev)	LRI	LRS	
1				LRI2	LRS1	LRS2
2	T1	S1	f2	LRI1	LRS1	LRS2
3				LRI3	LRS1	LRS2
4				LRI3	LRS1	LRS2
5	T1	S1	f1	LRI2	LRS1	LRS2
6				LRI1	LRS1	LRS2
7				LRI2	LRS1	LRS2
8	T1	S2	f1	LRI1	LRS1	LRS2
9				LRI3	LRS1	LRS2
10				LRI3	LRS1	LRS2
11	T2	S1	f1	LRI2	LRS1	LRS2
12				LRI1	LRS1	LRS2
13				LRI3	LRS1	LRS2
14	T2	S2	f2	LRI2	LRS1	LRS2
15				LRI1	LRS1	LRS2
16				LRI3	LRS1	LRS2
17	T1	S2	f2	LRI1	LRS1	LRS2
18				LRI2	LRS1	LRS2
19				LRI1	LRS1	LRS2
20	T2	S1	f2	LRI2	LRS1	LRS2
21				LRI3	LRS1	LRS2
22				LRI1	LRS1	LRS2
23	T2	S2	f1	LRI2	LRS1	LRS2
24				LRI3	LRS1	LRS2

of magnesium alloy UNS M11917, as indicated in Figure 4.27. This type of workpiece was chosen since it allows cylindrical and frontal milling operations, and it has good accessibility for obtaining measurements with the roughness tester, besides being easier to manufacture than the trapezoidal section. The combination of materials was chosen for the variety of magnesium alloy parts that have steel inserts (such as threaded bushes for fastening studs or screws), which can be machined at the same time as the part that contains them, when made on said parts repair and maintenance operations. Table 4.13 summarizes the results obtained during the milling trials.

4.4.4 Results: analysis and discussion

Once the roughness values *Ra* obtained during milling trials are collected, a fixed-effects analysis of variance (ANOVA) was conducted over them in order

(a)

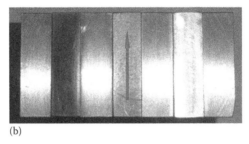

(b)

Figure 4.27 (a) Parts where the milling trials were carried out. (b) Machining direction.

to identify the most influential factors in the surface finish of magnesium-based hybrid workpieces. The initial model took into account all the interactions of order 2 and 3, and in each step, the factor with a higher p value—always greater than 0.05—was eliminated until all those that remained in the model were statistically significant. Because the hypotheses of normality and homoscedasticity were not met in the residuals of the model obtained with the original variable Ra, it was necessary to perform a logarithmic transformation on the response variable Ra. Table 4.14 shows the final result of the ANOVA analysis. The statistical analysis of the roughness values Ra of the milling trials was carried out using SAS software (SAS Institute 2004).

The results of the ANOVA analysis in Table 4.14 indicate that the spindle speed S, the location with respect to the insert LRI, and the interaction of these two variables $S*LRI$ have a statistically significant influence in the surface roughness of the workpieces of magnesium alloy UNS M11917 and steel insert UNS G11170 milled. In this machining process, neither the feed rate, the type of tool, nor the relative positions of the measurement with respect to the workpiece have a statistically significant influence. To be more precise, 39.12% of the variability in the roughness is due to the relative location with respect to the insert (which represents 58.57% of the controlled variability), 16.42% is due to the interaction of the spindle speed with the location with respect to the insert, and 11.25% is due to the spindle speed (see Table 4.15). The fact that the spindle speed significantly

Table 4.13 Values of the arithmetic mean deviation roughness, *Ra* (μm),
obtained during milling trials

Obs.	T	S (rpm)	f (mm/rev)	LRI	Ra (μm) LRS1	Ra (μm) LRS2
1				LRI2	1.50	0.83
2	T1	S1	f2	LRI1	0.12	0.17
3				LRI3	0.08	0.08
4				LRI3	0.17	0.35
5	T1	S1	f1	LRI2	1.13	0.67
6				LRI1	0.22	0.51
7				LRI2	0.29	0.21
8	T1	S2	f1	LRI1	0.12	0.09
9				LRI3	0.17	0.03
10				LRI3	0.10	0.12
11	T2	S1	f1	LRI2	0.69	0.64
12				LRI1	0.11	0.09
13				LRI3	0.10	0.12
14	T2	S2	f2	LRI2	0.14	0.20
15				LRI1	0.09	0.22
16				LRI3	0.34	0.15
17	T1	S2	f2	LRI1	0.21	0.13
18				LRI2	0.20	0.19
19				LRI1	0.12	0.16
20	T2	S1	f2	LRI2	0.46	1.21
21				LRI3	0.23	0.11
22				LRI1	0.12	0.30
23	T2	S2	f1	LRI2	0.20	0.17
24				LRI3	0.17	0.13

Table 4.14 Results of the ANOVA *Ra* Neperian logarithm

Source	DF	Sum of squares	Mean square	F	Pr > F
S	1	3.417	3.417	14.23	<0.001
LRI	2	11.883	5.941	24.73	<0.001
S*LRI	2	4.989	2.495	10.39	<0.001
Error	42	10.089	0.240		
Total	47	30.378			

Table 4.15 Variability percentage
explained by each factor

Source	Contribution percentage (%)
S	11.25
LRI	39.12
S*LRI	16.42

influences the milling of magnesium-steel hybrid parts is in line with various works, such as that of Sanz and coauthors (2008).

The reason why the location respect of the insert is the most determining parameter can be found in the different machining characteristics presented by steel (insert) and magnesium (rest of the workpiece), since the roughness obtained in the insert is clearly superior to that obtained before and after it. This fact can be seen in Figure 4.28, which, along with Figure 4.29, illustrates the dispersion of the Naperian logarithm of the roughness (ln Ra) with respect to the main influential factors (location with respect to the insert and spindle speed), considered in isolation. On the other hand, in general, when increasing the spindle speed, a slight reduction in the dispersion of the roughness is observed (see Figure 4.29).

The only interaction that affects the variable ln Ra in the model of the milling is $S*LRI$, such that when increasing the spindle speed, the surface finish in the insert ($LRI2$) improves considerably. Before the insert ($LRI1$) and after the insert ($LRI3$), no significant differences were observed. All this can be observed in the graph of Figure 4.30.

Equation 4.3 models the variability of ln Ra obtained from the ANOVA analysis for the milling tests, where μ is a constant term of adjustment to the mean of the values of ln Ra and α_i, β_j, and $\alpha\beta_{ij}$ represent the effect of the levels of the spindle speed, the location with respect to the insert, and the

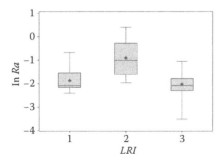

Figure 4.28 Box and whisker plot with dispersion of ln Ra versus location respect of the insert factor.

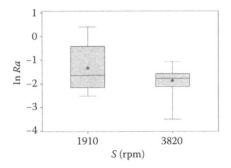

Figure 4.29 Box and whisker plot with dispersion of ln *Ra* versus spindle speed factor.

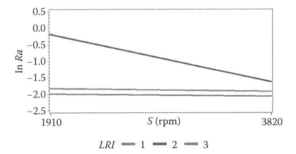

Figure 4.30 Interaction between spindle speed and location with respect to the insert for ln *Ra*.

interaction of spindle speed with the location with respect to the insert, respectively.

$$\ln(Ra_{ij}) = \mu + \alpha_i + \beta_j + \alpha\beta_{ij} + \varepsilon_{ij} \tag{4.3}$$

The estimations of the parameters of the model in Equation 4.3 are shown in Table 4.16. The hypotheses of this model were contrasted, verifying the normality hypothesis (Table 4.17), the homoscedasticity hypothesis, and the nonexistence of patterns in the model (Figure 4.31), fulfilling all the hypotheses.

From the obtained results, the variability of the surface roughness in milling can be modeled through Equation 4.4.

$$Ra_{ij} = \exp(\mu + \alpha_i + \beta_j + \alpha\beta_{ij} + \varepsilon_{ij}) \tag{4.4}$$

Table 4.16 Parameter estimates of the model of Equation 4.3

Parameter		Estimation	Standard error	t value	Pr > |t|
Intercept	μ	−1.9966	0.1733	−11.52	<0.0001
3820 rpm	α_1	−0.0646	0.2451	−0.26	0.7935
1910 rpm	α_2	0	–	–	–
Before the insert	β_1	0.1657	0.2451	0.68	0.5027
On the insert	β_2	1.8138	0.2451	7.40	<0.0001
After the insert	β_3	0	–	–	–
3820 rpm—before the insert	$\alpha\beta_{11}$	−0.0264	0.3466	−0.08	0.9396
3820 rpm—on the insert	$\alpha\beta_{12}$	−1.3809	0.3466	−3.98	0.0003
3820 rpm—after the insert	$\alpha\beta_{13}$	0	–	–	–
1910 rpm—before the insert	$\alpha\beta_{21}$	0	–	–	–
1910 rpm—on the insert	$\alpha\beta_{22}$	0	–	–	–
1910 rpm—after the insert	$\alpha\beta_{23}$	0	–	–	–

Table 4.17 Normality tests for the residuals of the model of Equation 4.3

Test	Statistic		p Value	
Kolmogorov-Smirnov	D	0.1087	Pr > D	>0.150
Cramer-von Mises	W-Sq	0.0845	Pr > W-Sq	0.183
Anderson-Darling	A-Sq	0.5841	Pr > A-Sq	0.127

With the model in Equation 4.4 and the parameter estimates (Table 4.16), the roughness values predicted by the model are obtained for the different combinations of the levels of the statistically influential parameters in the surface finish of the milling process. The ranking for these combinations, based on the roughness predictions obtained, is shown in Table 4.18.

From the obtained ranking, it is observed that the surface finish in milling operations is optimized with a spindle speed of 3820 rpm, and the value of *Ra* in the whole piece machining at that speed is expected to be minimized, which is the highest among those tested in the experiment. With that spindle speed, reaching quite good roughness levels, around 0.20 μm on the insert and around 0.14 μm in the rest of the

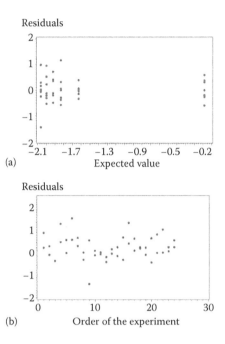

(a)

(b)

Figure 4.31 Checking the homoscedasticity hypothesis (a) and the nonexistence of patterns (b) in the model of Equation 4.3.

Table 4.18 Ranking of the combinations of the cutting conditions

S (rpm)	LRI	Ra (μm) estimated
3820	After the insert	0.13
1910	After the insert	0.14
3820	Before the insert	0.15
1910	Before the insert	0.16
3820	On the insert	0.20
1910	On the insert	0.83

machined surface, is expected. Furthermore, it is clearly shown that, in order to achieve a good surface quality on the steel insert, it is not advisable machining at a low speed. This result is in agreement with that obtained by Sanz and coauthors (2008). It should be noted that the results obtained constitute a flexible optimization of the milling process, allowing some freedom of choice both in the cutting tool and in the value of the feed rate.

4.5 Conclusions

This work shows the design and manufacturing of magnesium-based hybrid parts, which allow analyzing their behavior in these types of processes. Additionally, as an example of their possible use, an experimental application has been carried out with the milling workpieces. Specifically, typical repair and maintenance operations of the aeronautical industry carried out under low speed conditions have been simulated.

After the design and manufacture of the workpieces, it is possible to conclude that:

- The design of hybrid parts workpieces that allow performing machining trials and comparing of the results in a systematic way is necessary to continue advancing in the study on the behavior of hybrid parts, in particular, of the magnesium base. The choice of a design that allows fabrication as simple as possible and in the shortest time is crucial to obtain workpieces at the lowest cost.
- The use of adhesives implies an additional cost, but it is considered appropriate as part of the design to minimize the potential problems that may arise from the different types of workpieces during machining, in addition, it can protect the workpieces from galvanic corrosion.
- The manufacture of the workpieces is conditioned by the initial dimensions of the raw materials, and the difficulty and time of manufacture of the different components can be greatly reduced if they can be found with the final dimensions or so near them as possible.
- During the manufacture of the workpieces, it was possible to verify that:
  - For workpieces of turning, it can be stated that for any combination of materials (aluminium–magnesium, steel–magnesium, and titanium–magnesium), the rectangular section insert is the easiest to perform and the one with the lowest manufacturing time. The aluminium–magnesium and titanium–magnesium workpieces could also be made with inserts of circular section since the manufacturing and assembly times do not differ much from those obtained for the workpieces with inserts of a rectangular section. The workpieces with inserts of trapezoidal section are the ones that require more time of manufacture and entail greater difficulty for the adjustment of the inserts to the base; however, its use is not ruled out until its dynamic behavior is verified under high-speed conditions.
  - In the case of the milling workpieces, the simplest to manufacture and the ones that require less time are those with inserts with a circular section. These are followed, in increasing order

in difficulty and manufacture time, by the rectangular and trapezoidal inserts. In order to carry out tests of frontal milling and
cylindrical milling in the same position of the workpiece, the
rectangular section insert is the most suitable.

– In the case of drilling workpieces, as all have the same geometry,
  the differences in the manufacturing difficulty and the time
  invested in obtaining them are associated with the machinability of the materials of the respective inserts in the manufacture of
  the parallelepiped as in the previous drilling. Thus, aluminium
  is the one that has the least difficulty and requires lesser manufacturing time, followed by steel and titanium, respectively. It
  should also be noted, as a relevant conclusion, that for drilling
  operations, the workpieces with removable joints are very useful
  when it comes to easily accessing the measurement areas of the
  surface roughness.

• The experimental application has allowed optimizing the conditions
  of cut (feed rate, spindle speed, and type of tool) and obtaining a predictive model of surface roughness, in terms of average arithmetic
  roughness, *Ra*, as well as the selection of the combination of cutting
  conditions that best meets the quality requirements demanded in
  the automotive or aeronautical sector.

• In the milling tests carried out (face milling) on the UNS M11917
  magnesium-based hybrid workpiece with rectangular steel inserts
  UNS G11170, it was found that:
  – The feed rate is not statistically significant in the surface finish.
  – When increasing the spindle speed, an improvement in the surface
    roughness is observed.
  – The analysis of the interaction of spindle speed location with
    respect to the insert allows observing how, when increasing the
    spindle speed, there is a considerable improvement in the sur-
    face roughness in the steel insert, maintaining the roughness in
    the magnesium (before and after the insert).

• In the tests carried out, it was possible to confirm that the differences
  observed between measuring the roughness at the beginning of the
  workpiece and measuring it at the end of the workpiece are not statistically significant.

  For practical purposes, the following conclusions are also provided as
  a result of the interpretation of the conclusions drawn directly from
  the design and manufacture of the workpieces and of the experimental study that has been carried out with some of them, in addition to
  other technological, economic, and environmental factors:

• In face milling operations of UNS M11917 magnesium-based hybrid
  workpieces with rectangular inserts of steel UNS G11170, taking into
  account the ranking of combinations of cutting conditions based on

the estimated values of *Ra* through the predictive model obtained, it is possible to affirm that higher and lower values of spindle speed can be used since, even with these variations, the expected roughness still remains within the limits of the aeronautical and automotive sectors that are being analyzed.

* In this type of milling, the feed rate is not decisive for the roughness but decisive for the machining time; so, it is advisable to use the highest possible advances to reduce the time of machining and, thus, the associated costs.

Acknowledgments

The authors thank the following: the Research Group of the UNED "Industrial Production and Manufacturing Engineering (IPME)," who gave support during the development of this work; the Industrial Engineering School-UNED (Project REF2018-ICF05), who funded and is supported, in part, by four grants from the Ministerio de Ciencia e Innovación, Ministerio de Economía y Competitividad, Agencia Estatal de Investigación (AEI), and Fondo Europeo de Desarrollo Regional (FEDER) (DPI2014-58007-R, MTM2016-78227-C2-1-P, MTM2015-69323-REDT, CGL2014-58322-R), Spain; and the Antolín Group, for the material provided.

References

Aghion, E., Brontin, B., Von Buch, F., Schumann, S., Friedrich, H. 2003. Newly developed magnesium alloys for powertrain applications. *Journal of the Minerals, Metals & Materials Society* 55(11): 30–33.

Amancio, S. 2012. Innovative solid-state spot joining methods for fiber composites and metal-polymer hybrid structures. *Joining in Car Body*: 1–47.

Ashby, M.F., Brécht, Y.J.M. 2003. Designing hybrid materials. *Acta Materalia* 51: 5801–5821.

ASME B46.1. 2009. *Surface Texture (Surface Roughness and Lay)*. The American Society of Mechanical Engineers, New York.

Ben-Artzy, A., Munitz, A., Kohn, G., Bronfin, B., Shtechman, A. 2002. Joining of light hybrid constructions made of magnesium and aluminum alloys. *Magnesium Technology 2002, TMS (The Minerals, Metals & Materials Society)*: 295–302.

Carro, J. 1998. *Curso de Metrología Dimensional*. Escuela Técnica Superior de Ingenieros Industriales de Madrid, Madrid.

Casalino, G. 2017 Advances in welding metal alloys: Dissimilar metals and additively manufactured parts. *Metals* 7(2): 32–36.

Casalino, G., Guglielmi, P., Lorusso, V.D., Mortello, M., Peyre, P., Sorgente, D. 2017. Laser offset welding of AZ31B magnesium alloy to 316 stainless steel. *Journal of Materials Processing Technology* 242: 49–59.

Dau, J., Lauter, C., Damerow, U., Homberg, W., Tröster T. 2011. Multi-material systems for tailored automotive structural components. *Proceedings of the 18th International Conference on Composite Materials*, Jeju Island, South Korea.

DRL. 2011. Innovation Report. Institute of Composites Structures and Adaptive Systems, Braunschweig.

Fischersworring-bunk, A., Landerl, C., Fent, A., Wolf, J. 2006. The new BMW inline six-cylinder composite Mg/Al crankcase. *IMA, Annual World Magnesium Conference* 62: 49–58.

Frantz, M., Lauter, C., Tröster T. 2011. Advanced manufacturing technologies for automotive structures in multi-material design consisting of high-strength steels and CFRP. *Proceedings of the 56th International Scientific Colloquium*, Ilmenau, Germany.

Gururaja, M.N., HariRao, A.N. 2012. A review on recent applications and future prospectus of hybrid composites. *International Journal of Soft Computing and Engineering* 1(6): 352–355.

Henkel. 2011. *Soluciones para la industria. Adhesivos para la industria, selladores y productos para el tratamiento de superficies.* Henkel AG & Co. Madrid (Spain)

ISO 8688-1. 1989. *Tool Life Testing in Milling—Part 1: Face Milling*. International Organization for Standardization, Geneva, Switzerland.

ISO 8688-2. 1989. *Tool Life Testing in Milling—Part 2: End Milling*. International Organization for Standardization, Geneva, Switzerland.

ISO 4287. 1997. *Geometrical Product Specifications (GPS)—Surface Texture: Profile Method—Terms, Definitions and Surface Texture Parameters*. International Organization for Standardization, Geneva, Switzerland.

ISO 3685. 1998. *Tool-Life Testing with Single-Point Turning Tools*. International Organization for Standardization, Geneva, Switzerland.

ISO 4288. 1998. *Geometrical Product Specifications (GPS)—Surface Texture: Profile Method. Rules and Procedures for the Assessment of Surface Texture*. International Organization for Standardization, Geneva, Switzerland.

ISO 1302. 2002. *Geometrical Product Specifications (GPS)—Indication of Surface Texture in Technical Product Documentation*. International Organization for Standardization, Geneva, Switzerland.

ISO 377. 2013. *Steel and Steel Products—Location and Preparation of Samples and Test Pieces for Mechanical Testing*. International Organization for Standardization, Geneva, Switzerland.

ISO 10123. 2013. *Adhesives—Determination of Shear Strength of Anaerobic Adhesives Using Pin-and-Collar Specimens*. International Organization for Standardization, Geneva, Switzerland.

ISO 243. 2014. *Turning Tools with Carbide Tips—External Tools*. International Organization for Standardization, Geneva, Switzerland.

ISO 514. 2014. *Turning Tools with Carbide Tips—Internal Tools*. International Organization for Standardization, Geneva, Switzerland.

Kalpakjian, S., Schmid. S.R. 2008. *Manufactura, ingeniería y tecnología*. Prentice Hall, México.

Kickelbick, G. 2007. *Hybrid Materials. Synthesis, Characterization and Applications*. Wiley-VCH Verlag GmbH & Co., Darmstadt.

Krebs, R., Böhme, J., Doerr, J., Rothe, A., Schneider, W., Haberling, C. 2005. Magnesium. Hybrid turbo engine from Audi. *MTZ Worldwide* 66(4): 13–16.

Lee, D., Morillo, C., Oller, S., Bugeda, G., Oñate, E. 2013. Robust design optimization of advance hybrid (fiber–metal) composite structures. *Composite Structures* 99: 181–192.

Luo, J.G., Acoff, V.L. 2000. Interfacial reactions of titanium and aluminum during diffusion welding. *Welding Journal* 79(9): 239s–243s.

Manikandan, G., Uthayakumar, M., Aravindan, S. 2012. Machining and simulation studies of bimetallic pistons. *International Journal of Advanced Manufacturing Technology: Technology* 66(5–8): 711–720.

Mg Showcase issue 1. 2007. International Magnesium Association, May.

Montgomery, D.C. 2012. *Design and Analysis of Experiments.* John Wiley & Sons, New York.

Nasiri, A.M., Li, L., Kim, S.H., Zhou, Y., Weckman, D.C., Nguyen, T.C. 2011. Microstructure and properties of laser brazed magnesium to coated steel. *Welding Journal* 90: 211–219.

Nasiri, A.M., Weckman, D.C., Zhou, Y. 2013. Interfacial microstructure of diode laser brazed AZ31B magnesium to steel sheet using a nickel interlayer. *Welding Journal* 92: 1–10.

Perez, J.M. 1998. Tecnología mecánica 1. Escuela Técnica Superior de Ingenieros Industriales de Madrid, Universidad Politécnica de Madrid, Madrid.

Qehaja, N., Salihn, A., Zeqiri, H., Osmani, H., Zeqiri, F. 2012. Machinability of metals, methods and practical application, *Annals of DAAAM for 2012 & Proceedings of the 23rd International DAAAM Symposium* 23(1): 29–32.

Qi, X., Song, G. 2010. Interfacial structure of the joins between magnesium alloy and mild steel with nickel as interlayer by hybrid laser-TIG welding. *Materials and Design* 31: 605–609.

Riba, C. 2008. *Selección de materiales en el diseño de máquinas.* Ediciones Universidad Politécnica de Cataluña, Barcelona.

Rubio, E.M., Camacho, A.M., Sánchez, J.M., Marcos, M. 2005. Surface roughness of AA7050 alloy turned bars, analysis of the influence of the length of machining. *Journal of Materials Processing Technology* 162–163: 682–89.

Ruiz, P.A. 2011. Adhesivos estructurales: Alternativa potente y eficaz para la unión de metales. *Revista Metal Actual* 20: 60–67.

Saénz de Pipaón, J.M. 2013. *Diseño y fabricación de probetas de componentes híbridos con aleaciones de magnesio para ensayos de mecanizado.* Department of Manufacturing Engineering, Industrial Engineering School, UNED, Madrid.

Sanz, C., Fuentes, E., Gonzalo, O., Bengoetxea, I., Obermair, F., Eidenhammer, M. 2008. Advances in the ecological machining of magnesium and magnesium-based hybrid parts. *International Journal of Machining and Machinability of Materials* 4(4): 302–319.

SAS Institute. 2004. *SAS/Stat User's guide.* Release 9.1. SAS Institute, Cary, NC.

Seco. 2012. *Catálogo y guía técnica* (secotools.com).

Staeves, J. 2005. Zukünftige magnesium-auwendungen im automobilbereich, *BMW Group, 13th Magnesium Automotive and End User Seminar,* Aalen, 1–16.

Suryawanshi, B.K., Prajitsen, G.D. 2013. Review of design of hybrid aluminum/composite drive shaft for automobile. *International Journal of Innovative Technology and Exploring Engineering* 2(4): 259–266.

Tharumarajah, A., Koltum, P. 2007. Is there an environmental advantage of using magnesium components for light-weighting cars? *Journal of Cleaner Production* 15: 1007–1013.

UNE 16148. 1985. *Ensayos de duración de herramientas de torno de corte único.* Asociación Española de Normalización y Certificación, Madrid.

UNE 36423. 1990. *Ensayos de acero. Determinación del índice de maquinabilidad. Métodos de torneado de corta duración.* Asociación Española de Normalización y Certificación, Madrid.

Uthayakumar, M., Prabhaharan, G., Aravindan, S., Sivaprasad, J.V. 2008. Machining studies on bimetallic pistons with CBN tool using the Taguchi method-technical communication. *Machining Science and Technology* 12(2): 249–255.

Villeta, M., de Agustina, B., Sáenz de Pipaón, J.M., Rubio, E.M. 2012. Efficient optimisation of machining processes based on technical specifications for surface roughness: Application to magnesium pieces in the aerospace industry, *The International Journal of Advanced Manufacturing Technology* 60, 9–12: 1237–1246.

Wagner, F., Zerner, I., Kreimeyer, M., Seefeld, T., Sepold, G. 2001. Characterization and properties of dissimilar metal combinations of Fe/Al and Ti/Al sheet materials. *Proceedings ICALEO'01 (CD-ROM)*, Laser Institute of America, Orlando.

Yan, L., Cuirong, L., Haibo, Y., Fei, Z., Zhisheng, W. 2017. Numerical simulation of Ti/Al bimetal composite fabricated by explosive welding. *Metals* 7(10): 407–419.

Zakaria, B., Yazid, H. 2017. Friction stir welding of dissimilar materials aluminum Al6061-T6 to ultra low carbon steel. *Metals* 7(2): 42–50.

chapter five

Laser surface processing of magnesium alloys

Yingchun Guan, Shuquan Zhang, Jia Li,
Xiangjun Tian, and Hongyu Zheng

Contents

5.1 Introduction

Mg alloys exhibit excellent specific strength and stiffness, good damping capacity, and high thermal conductivity and machinability, which have promising applications in the transport industry, electronics industry, and aircraft-space industry. However, low ductility of Mg alloys due to their crystal structural characteristics and low ignition point of the machined chips present challenges of overheating and burning during machining processes. Investigating and reducing inflammability have attracted considerable research interest. Use of flushing coolants can effectively prevent Mg ignition but results in the pollution of environment and makes reclamation of the machined chips a tricky issue. One alternative approach is dry machining without the use of flushing coolant. Through process optimization, Mg ignition can be avoided. Hou et al. (2010) performed a series of studies in cutting various magnesium alloys (including AM50A and AZ91D). The effect of cutting parameters (such as cutting speed, feed rate, and depth of cut) on ignition of chips in face milling process was investigated. It was reported that for a given cut depth, the ignition of fire in the forms of sparks, flares, or ring occurred only in moderate cutting speeds or feed rates and can thus be prevented by changing these parameters. In order to understand the mechanisms of Mg chip ignition in machining, the chips produced in different machining processes were collected and analyzed with regard to the surface morphology changes. Mg alloys with considerable amount of secondary phases exhibit different machining characteristics. Sunil et al. (2016) studied two AZ series Mg alloys, i.e., AZ31 and AZ91, to assess how the amount and distribution of the secondary phases would affect the machining characteristics. The drilling process was performed using different sets of process parameters and varying cutting forces. Chips produced in the drilling process were analyzed. From the results, it appears clear that the presence of secondary phase (Mg17Al12) has a significant influence on the cutting forces. The increase in cutting speed has reduced the required cutting force and load fluctuations in all the cases.

Machining of Mg alloys under different conditions has been reported. Carou et al. (2014) carried out an experiment concerning intermittent turning of UNS M11917 Mg alloy with the use of dry machining and minimum quantity lubrication system and analyzed different machining conditions by controlling the cutting speed, depth of cut, and feed rate. To evaluate intermittent turning process, continuous bars and slotted bars were used. The process was evaluated by taking the surface roughness as the response variable. Full factorial experimental designs were adopted and their results were analyzed by using analysis of variance. The identification of feed rate is one main result of statistical analysis and a significant factor for all the tests, explaining the most part of the analyzed variability. In contrast, it was found that both the cutting speed and type of interruption were not the significant sources of variability when analyzed in isolation. Furthermore, surface roughness, in terms of Ra, was identified to be more dispersed when machining at low feed rates. The depth of cut and its interaction with feed rate are also found to be important sources of variability in the surface roughness.

In this chapter, we report research development and advances in nonconventional machining, i.e., short pulse laser processing of Mg alloys in the respective sections.

5.2   Machining of magnesium alloy

Laser machining has unique characteristics, including noncontact (thus no tool wear), controlled heat-affected zone, and the possibility of enhancing machined surface properties. In this section, we present recent research work on laser processing of pure Mg, Mg alloys AZ91D and AZ31B, and Mg-RE alloy GW103K. Material interactions with short pulse lasers, including nanosecond and femtosecond, are described in details. First, the microstructure evolution of AZ91D Mg alloy surface caused by both millisecond and nanosecond pulse Nd:YAG laser melting is introduced and compared. In particular, local boiling and superheating phenomenon are discussed with in-depth analysis. At appropriate processing conditions, laser beam energy causes surface oxidation on AZ31B Mg alloy and leads to surface coloration probably due to light interference of the incident laser beam and the reflected laser beam. A study of surface coloration by both nanosecond Nd:YAG laser and excimer laser irradiation is described in this section. Second, femtosecond laser interaction with materials generates periodical surface structures that correspond to different surface colors. This section aims to address issues of how these periodical surface structures are formed, what the correlation between these structures and the surface colors is, and what the potential applications of the surface structures are on pure Mg and the two types of Mg alloys. Finally, the microstructure and mechanical property of the laser melting deposition (LMD)-produced GW103K alloy is described before and after conventional heat treatment, where a continuous wave CO_2 laser with a power of 4 kW is used for LMD processing. The low pressure sand cast GW103K alloy is especially used for comparison with LMD-produced alloy in this section. LMD-produced GW103K alloy consists of fine equiaxed grain (19 μm) structure and small Mg_3Gd and GdH_2 particles. Compared with sand cast GW103K alloy, LMDed alloy shows less improvement in yield strength (YS) and ultimate tensile strength (UTS) after T6 treatment (solution plus aging treatments), while it still has a significantly higher elongation (+5.8%).

5.2.1   Nanosecond pulse laser machining of Mg alloys

Short pulse lasers have high peak power that raises material surface temperature beyond its boiling point and directly affects the surface features and microstructures of the irradiated area (Bhattacharya et al., 1991, Craciun et al., 2002, Jacobs and Rikken, 1987, Rubahn, 1999). Jacob and Rikken (1987) studied laser-induced local boiling and observed the generation of cavitation bubbles at the solid–liquid interface. The morphology of laser-irradiated surface changed significantly as compared to the original surface due to the local boiling effects during the laser irradiation. Subsurface integrity is also an important issue in surface machining. It was reported that subsurface superheating has been observed in nanosecond laser irradiation (Bhattacharya et al., 1991, Craciun et al., 2002, Rubahn, 1999, Tillack et al., 2004, Steen, 2003) due to the energy transfer to beneath the surface. Under certain conditions, the temperature beneath the surface was observed higher than the top surface temperature probably due to faster latent heat cooling at the surface. Such phenomena is called subsurface superheating. The high internal temperatures contribute to the nucleation of the gaseous phase at the near surface, therefore resulting in "microexplosions" of the material during laser irradiation and, correspondingly, volume expulsion (Craciun et al., 2002). However, little public literature is available on how short-pulse lasers affect the surface structures and the solidification microstructures of the Mg alloys such as AZ91D. In this section, we discuss the kinetics of rapid solidification and phase transformation as well as the mechanism of boiling effect and crater formation during laser irradiation on AZ91D Mg alloy.

Figure 5.1 illustrates the interaction between nanosecond pulse Nd:YAG laser irradiation and AZ91D Mg alloy in a sealed chamber with argon protection (Guan et al., 2014). The surface profile of craters was further studied using stylus profilometer, as shown in Figure 5.2 (Guan et al., 2014). It is seen that the diameter of typical big craters is nearly 50 μm. A typical crater has a shape of conical profile with the ripples-structure distributing on wall surface of the hole. In addition, molten materials surrounding the craters were also observed in the liquid layer at the surface. Meanwhile, the average depth of the craters was found to be 15 μm according to the profile of the craters.

Figure 5.3 illustrates the cross-sections of AZ91D Mg alloy after laser treatment (Guan et al., 2014). It is shown in Figure 5.3a that the thickness of the melted layer is about 20 μm, which is much smaller than 150 μm— results reported on millisecond laser-induced surface melting (Guan et al., 2009a, 2009b, 2009c, 2010a, 2010b, Zhou et al., 2007). In Figure 3.3b, molten materials inside the crater are observed. Moreover, the size of the crater is consistent with the results in Figure 5.2. However, no cellular/dendrite solidification microstructure was found in the melt layer. Furthermore, quantitative analyses of chemical contents showed that molten materials contained an Al content of 10.2–11.1 wt%, which was obviously higher than the average Al content of 9.0 wt% in the as-received material.

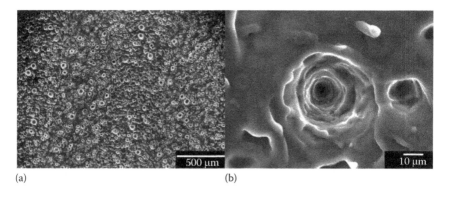

(a)   (b)

Figure 5.1 Typical SEM images of AZ91D Mg alloy surface after nanosecond laser irradiating: (a) lots of craters and (b) high magnification of the craters in (a). (From Guan, Y.C., Zhou, W., Li, Z.L., Zheng, H.Y., *Surf. Coat. Technol.*, 252, 168–172, 2014. With permission.)

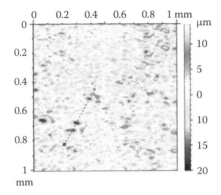

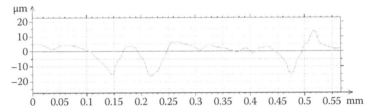

Figure 5.2 Surface heights of typical craters (bottom) measured along the path shown as the dotted line (upper) at the AZ91D Mg alloy surface after nanosecond laser irradiating. (From Guan, Y.C., Zhou, W., Li, Z.L., Zheng, H.Y., *Surf. Coat. Technol.*, 252, 168–172, 2014. With permission.)

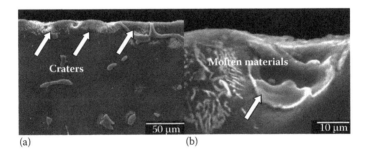

Figure 5.3 SEM images showing the cross-sections of AZ91D Mg alloy surface after nanosecond laser irradiating: (a) craters formed in melted layer and (b) crater with high magnification of (a). (From Guan, Y.C., Zhou, W., Li, Z.L., Zheng, H.Y., *Surf. Coat. Technol.*, 252, 168–172, 2014. With permission.)

This could be relatively caused by more Mg vaporization behavior in laser melting (Guan et al., 2010a, 2010b, Zhou et al., 2007). Based on the experiments, the various craters and solidification microstructure with higher Al concentration in the laser treated layer has no positive effect on surface properties of laser irradiated AZ91D Mg alloy.

Thermal effect during laser irradiation of metals is reported to be a main factor resulting in the precision and quality of the surface structures. In the nanosecond pulse laser system, assume that the thermalization in the electron subsystem is very fast; therefore, the heat transport inside the material can be calculated from a one-dimensional model (Peercy, 1986):

$$C_e \frac{\partial T_e}{\partial t} = k_e \nabla^2 T_c - g(T_e - T_l) + S; \tag{5.1}$$

$$C_l \frac{\partial T_l}{\partial t} = g(T_e - T_l); \tag{5.2}$$

$$S = I(t)A\alpha \, \exp(-\alpha z) \tag{5.3}$$

where T_e and T_l are the electron and lattice temperatures, C_e and C_l are the heat capacities of the electron and lattice subsystems, $I(t)$ is the laser intensity, g is the constant of the electron-phonon coupling, S is the laser heating source, α is the absorption coefficient, A is the surface transmission, z is the coordinate axis at the direction perpendicular to the surface, and k_e is the electron thermal conductivity.

In Eqs. 5.1 through 5.3, a thermal conductivity for the lattice subsystem is negligible. Electrons can be heated to very high transient temperature due to the much less heat capacity than the lattice (Dou et al., 2001). T_e, T_l, and T_L are characteristic time constants in Eqs. 5.1 through 5.3. $T_e = C_e/g$ is the electron cooling time, $T_l = C_l/g$ is the lattice heating time, and T_L is the laser pulse duration. It should be noted that these two temperature analyses are mostly used for ultrashort pulse laser irradiation on metal. However, to the best of our knowledge, the valid heat transfer model after nanosecond pulse laser processing is missing. Therefore, it was assumed that $\tau_L \gg \tau_l$ and the electron temperature is equal to the lattice temperature $T_e = T_l = T$. Eqs. 5.1 through 5.3 can be summarized as follows:

$$C_e \frac{\partial T_e}{\partial t} = k_e \nabla^2 T_c + IA\alpha \, \exp(-\alpha z). \tag{5.4}$$

Therefore, energy loss in nanosecond laser irradiation should be attributed to heat diffusion, because it has enough time to thermalize optical energy and conduct heat (Steen, 2003). Further discussion about heat transfer model during nanosecond laser irradiation is needed for better understanding.

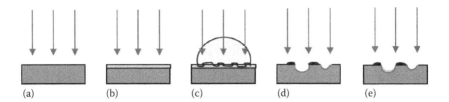

Figure 5.4 Schematic diagram illustrating the nanosecond laser-induced crater formation on target surface. (a) Laser energy deposition and heat conduction, (b) occurrence of surface melting and boiling, (c) generation of plasma expansion and shock waves, (d) cavitation bubble generation, and (e) bubble oscillation and crater formation. (From Guan, Y.C., Zhou, W., Li, Z.L., Zheng, H.Y., *Surf. Coat. Technol.*, 252, 168–172, 2014. With permission.)

Laser-induced crater formation is presented in Figure 5.4 (Guan et al., 2014). As nanosecond laser irradiates on Mg alloy, incident laser at high fluence (power density 1.37×10^8 W/cm^2) generates a large amount of heat on the target surface. Moreover, the accumulated energy causes the melt of material due to low melting point 595°C for AZ91D Mg alloy. Furthermore, as surface temperature increases to critical point, boiling behavior takes place (Rubahn, 1999, Steen, 2003), as shown in Figure 5.4a.

The motion of planar vaporization interface is sustained by local heat flow provided that (Bhattacharya et al., 1991)

$$L_v \frac{dS}{dt} = -K_s \left(\frac{\partial T_v}{\partial x} \right)_{x=S}, \tag{5.5}$$

where K_s is surface thermal conductivity of solid, T_v is vaporization temperature, L_v is latent vaporization heat, and dS/dt is the velocity of vaporizing interface. According to Eq. 5.5, it can be concluded that the target with large ratio of latent heat of vaporization to thermal conductivity increases the degree of subsurface superheating (Bhattacharya et al., 1991). For AZ91D Mg alloy, thermal conductivity was measured as 55 W/mK, and to the best of our knowledge, latent heat of vaporization has not been determined experimentally. However, Figure 5.2 and Figure 5.4 exhibit expulsion of the melted material from near surface of AZ91D Mg alloy. The melted matter observed at the surface is related to internal temperature being higher than evaporating surface temperature, so-called subsurface superheating. Therefore, we speculate that subsurface superheating occurs during laser irradiation.

Nucleation of gaseous phase at near surface may be occur due to the high subsurface superheating temperature, hence resulting in "micro-explosions" of target material at some stage in laser irradiating process.

These "microexplosions" lead to the volume expulsion of the material (see Figure 4 in Craciun et al., 2002). Therefore, we attribute the formation of craters and the generation of melted magnesium alloy to the laser-induced explosive volume boiling effect, as shown in Figure 5.4b. Furthermore, as subsequent incident laser pulses focus on the liquid material at boiling surface, optical breakdown occurs due to nonlinear absorption, and plasma formation is caused, as shown in Figure 5.4c.

Generally, it is accepted that plasma expansion occurs following the emission of shock waves and the formation of cavitation bubbles (Chen et al., 2009, Jacobs and Rikken, 1987). In general, plasma expands adiabatically at a supersonic velocity, resulting in high-pressure shock waves in front of it. As laser plasma cools, the front shock waves rapidly detach from plasma because the velocity of shock waves is much greater than the particles behind it. The laser-induced bubbles continue to expand outward until all its initial kinetic energy converts into potential energy. Subsequently, the bubbles begin to implode adiabatically under the external liquid pressure. Moreover, the volume expulsion-induced melted matter can act as "fresh" defects and cause a next preferential laser irradiation for craters formation, as shown in Figure 5.4d.

Both temperature and pressure inside the cavity rise again, leading to the rebound of the generated bubbles. Usually, a bubble oscillates for several cycles until the energy is depleted completely and the gases fully dissolve into the surrounding liquid, especially in multiple laser pulses irradiating process. Bubble oscillation causes the generation of conical profile, with ripples structures on the wall. As a consequence, a typical crater with an average depth of several microns is formed, as presented in Figure 5.2b and Figure 5.3. Finally, by this means, many gathered crater clusters were generated after several consecutive laser pulses, which leads to the formation of craters on surface with different crater size (Figure 5.2a), as shown in Figure 5.4e.

It should be noted that such craters could be also generated on the ingot pure Mg and wrought AZ31B Mg alloy by the same laser irradiating process, as shown in Figure 5.5 (Guan et al., 2014). However, the amount and size of craters decreased significantly compared to that of AZ91D Mg alloy. Therefore, we propose that alloying elements in Mg alloys influences the mechanism of crater formation during subsurface superheating and explosive volume boiling. Further effort for deep understanding is still needed.

5.2.2 *Femtosecond pulse laser processing of Mg and Mg alloy*

Femtosecond laser causes iridescent effect on AZ31B Mg alloy surface. Figure 5.6 illustrates the effect after 10 femtosecond laser pulses (Guan et al., 2013). The optical properties of laser-irradiated Mg alloy are found to be changed significantly, and the surface exhibits different visual colors at different viewing angles. It shall be noted that such iridescent effects have

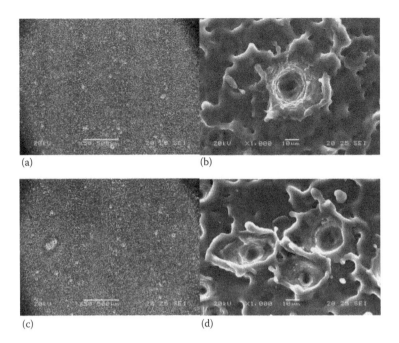

(a)   (b)

(c)   (d)

Figure 5.5 Typical SEM images of craters on the irradiated surface following nanosecond pulse Nd:YAG laser: (a) pure Mg, (b) close-up of (a), (c) AZ31B Mg alloy, and (d) close-up of (c). (From Guan, Y.C., Zhou, W., Li, Z.L., Zheng, H.Y., *Surf. Coat. Technol.*, 252, 168–172, 2014. With permission.)

Figure 5.6 Iridescent effect of the irradiated AZ31B Mg alloy surface after femto-second laser irradiating for 10 times. (From Guan, Y.C., Zhou, W., Li, Z.L., Zheng, H.Y., *J. Phys. D Appl. Phys.*, 46, 425305, 2013. With permission.)

also been observed on the irradiated surface of pure Mg and AZ91D Mg alloy by a similar femtosecond laser irradiating process. In this section, we focus on investigating AZ31B Mg alloy due to its wide practical uses and its predominantly homogeneous α-Mg microstructure.

Figure 5.7 shows typical surface reflectance before and after laser irradiation using a spectrophotometer (Guan et al., 2013). The appearance of noise signal in the wavelength range below 475 nm was attributed to the environment effect. The fine polished surface shows nearly 100% reflectance invisible spectral range from 475 to 750 nm. After laser irradiation, three dominate peaks appeared in the reflectance spectrum, corresponding to the wavelength of blue, green, and orange, respectively. Moreover, small wide peaks were also observed in the wavelength range of 570–600 nm, which corresponds to the color yellow. This agrees well with the observation in Figure 5.7. Furthermore, it was noted that the spectra pattern for all the laser-treated surfaces was the same, but reflectance decreased to 10%–70% with respect to the untreated surface as laser scan times increased from 10 to 50 times. The reduced reflectance at larger laser scan times was attributed to light trapping effect of the darken surface caused by longer laser irradiation time (Kannan et al., 2008, Zhang et al., 1997).

As shown in Figure 5.8 (Guan et al., 2013), scanning electron microscopy (SEM) was used to study the morphological evolution of the previous irradiated surfaces. When laser scan times were 10 times, fine ripples were produced above microsized wavy structures (grooves) on the surface, as illustrated in Figure 5.8a. Moreover, the orientation of ripples with the period in the range of 400–650 nm was found to be perpendicular to laser polarization direction, which was obviously less than the wavelength of incident light of 775 nm. Furthermore, the grooves' orientation with the period range around 1–3 μm

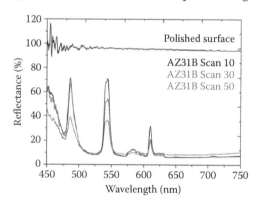

Figure 5.7 Reflectance of nonirradiated area at the polished surface and irradiated areas at the surface of AZ31B Mg alloy with progressive laser scan times. (From Guan, Y.C., Zhou, W., Li, Z.L., Zheng, H.Y., *J. Phys. D Appl. Phys.*, 46, 425305, 2013. With permission.)

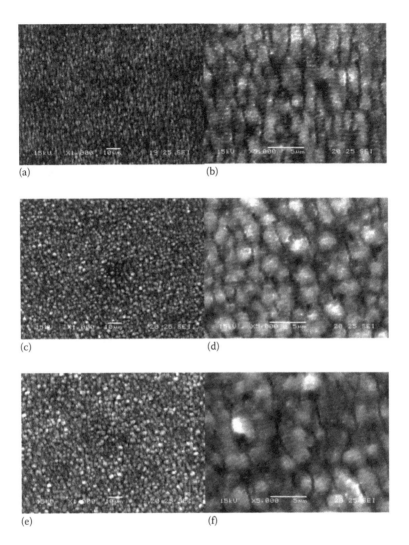

Figure 5.8 SEM images showing the morphological evolution of AZ31B Mg alloy surface after irradiation with progressively scan times: (a) 10 times, (b) magnified image for (a), (c) 30 times, (d) magnified image for (c), (e) 50 times, and (f) magnified image for (e). (From Guan, Y.C., Zhou, W., Li, Z.L., Zheng, H.Y., *J. Phys. D Appl. Phys.*, 46, 425305, 2013. With permission.)

was perpendicular to the ripple structures and parallel to laser polarization direction. Increasing the laser scan times to 30 times and 50 times, grooves expanded and cluster structures formed in the irradiated area due to surface melting under accumulation thermal effect. Moreover, the ripple orientation was similar to the previous results, but the period slightly decreased, as shown in Figure 5.8c–f. For laser scan times larger than 100 times, more surface melting took place at the whole surface, and the grooves and ripples disappeared gradually. It can be concluded here that the structures of grooves and ripples were highly depended on laser fluence. This is in accord with the finding of Bonse et al. (2005) (see Figure 6 in this reference). Unfortunately, further explanation of grooves formation was not given.

As illustrated in Figure 5.9 and Figure 5.10 (Guan et al., 2013), atomic force microscopy (AFM) was used for further investigation of surface topography at the irradiated areas. According to surface profile, the period range of ripples was found to be 450–620 nm, which is in accord with the SEM images. Moreover, when laser scan times were less, shallow groves and random nanoprotrusions were produced, as illustrated in Figure 5.9a and Figure 5.10a. Figure 5.9b shows that nanoripples replaced nanoprotrusions on the irradiated area with the increasing laser scan times. Figure 2.10b illustrates that the average of such ripples period is 530 nm. Figure 5.9d and Figure 5.10d illustrate that the average ripple period decreased to 450 nm with the laser scan times increasing to 50 times. Furthermore, grooves were identified as valleys in Figure 5.9, and they deepened and widened with the increasing laser scan

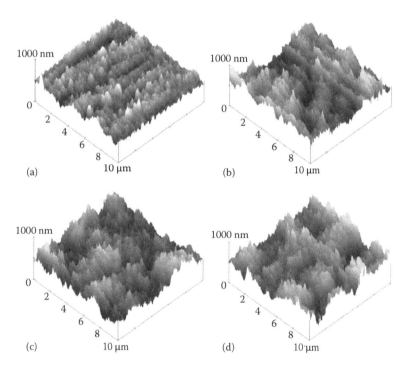

Figure 5.9 AFM images showing topographical evolution on AZ31B Mg alloy surfaces. The probed areas were 10 × 10 μm² for all the measurements: (a) scan 5 times, (b) scan 10 times, (c) scan 30 times, (d) scan 50 times. (From Guan, Y.C., Zhou, W., Li, Z.L., Zheng, H.Y., *J. Phys. D Appl. Phys.*, 46, 425305, 2013. With permission.)

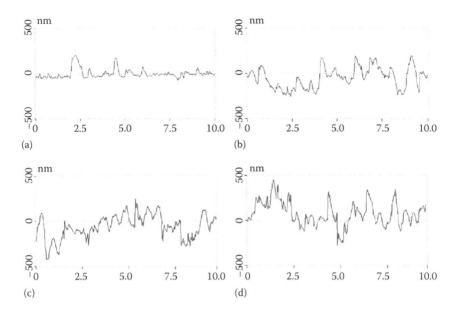

Figure 5.10 AFM images of surface profile measurement in Figure 5.8: (a) scan 10 times, (b) scan 10 times, (c) scan 30 times, and (d) scan 50 times. (From Guan, Y.C., Zhou, W., Li, Z.L., Zheng, H.Y., *J. Phys. D Appl. Phys.*, 46, 425305, 2013. With permission.)

times. Moreover, the surface roughness Ra of nonirradiated polished surface was 7.6 nm, which was much smoother than that of measurement area with 113.3 nm, 136.9 nm, 164.5 nm, and 187.5 nm, respectively.

Femtosecond laser-induced ripples and grooves on Mg alloy surface at near-damage-threshold fluence will be discussed briefly in the following. It is known that femtosecond laser irradiate on metal surface results in strong excitation. That could lead to species rapid ejection, such as atoms, photons, or electrons. Meanwhile, metal lattice will get energy from excited electrons by typical electron–phonon coupling. In the earlier laser pulses, nanoprotrusions randomly generated at the irradiated surface due to nonthermal melting as well as material redeposition effect within several picoseconds during rapid cooling condensation and plume collapse process (Li et al., 2009, Zheng et al., 2008). Laser-ejected species and plasma confinement during the irradiation process are the main causes of such nanoprotrusions (Itina et al., 2007, Schmidt et al., 2000). The temperature and the kinetic energy of ejected species dramatically increased with the increase in laser pulse number.

As shown in Figure 5.12a (Guan et al., 2014), the dispersing of the ejected species occurred along with the high energy expansion, which thereby led to the spreading of nanoprotrusions and so did the irradiated area filled with these protrusions. Meanwhile, the combination of incident laser with nano-protrusions would excite both localized and diffusing surface plasmons (SPs)

(Huang et al., 2009). The nanoprotrusions grew as the laser pulse number increased, and the SP was further excited by the evolved structures. Figure 5.12a–d and Figure 5.13a–d showed that SP would subsequently interfere with the incident laser wave, and this interference resulted in the formation of ripples (Guan et al., 2014). Furthermore, in consideration of orientation of ripples perpendicular to the laser polarization direction, it suggested to explain such ripples as the interference between the plasma wave electric field and the incident laser light field (Bonse et al., 2000, Pereira et al., 2004). However, the period of ripples in this study (480–600 nm) was found to be not following the classical relationship for ripples' subwavelength:

$$\Lambda = \frac{\lambda}{1 \pm \sin \theta}. \tag{5.6}$$

Taking into account the impact of SP, the revised scattering model can be transformed as follows (Huang et al., 2009):

$$\Lambda = \frac{\lambda}{\dfrac{\lambda}{\lambda_S} \pm \sin \theta} \tag{5.7}$$

Where λ_s is the wavelength of SP, and λ_s in the air/metal interface is given by

$$\lambda_S = \lambda \sqrt{\frac{\varepsilon_m + \varepsilon_d}{\varepsilon_m \varepsilon_d}}, \tag{5.8}$$

where $\varepsilon_m = \varepsilon_{m'} + {}_i\varepsilon_{m''}$ is the metal dielectric constant, and $\varepsilon_d = \varepsilon_{d'} + {}_i\varepsilon_{d''}$ is the dielectric medium dielectric constant. According to Eq. 5.7, in normal incidence ($\theta = 0°$), it can be obtained that the simple relationship $\Lambda = \lambda s$. In other words, the wavelength of interference fringes was equal to the period of induced ripples, which is always shorter than λ. This is consistent with the experimental results and is precisely fit to the generation of ripples' subwavelength (Figure 5.11 and Figure 5.13). In addition, it should be observed that the periods of ripples here have a uneven distribution in different spaces, and it indicated that both Mg alloy surfaces and air have different effective dielectric constants during the laser irradiation process, and this can have impact on the total effective refractive index for the SP, thus causing the different periods (Vorobyev and Guo, 2008).

Moreover, the average periods of ripples decreased with laser scan times slightly because of grating-assisted SP–laser combining mechanism (Bonse et al., 2000, Huang et al., 2009, Yang and Debroy, 1999). Our results

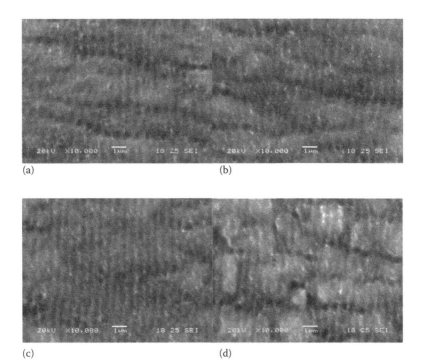

(a)   (b)

(c)   (d)

Figure 5.11 SEM images showing the evolution of two ripples on Mg surface after processing with progressively laser pulses: (a) 200 pulses, (b) 300 pulses, (c) 500 pulses, showing fine ripples above grooves, and (d) 500 pulses, showing fine ripples above cluster structures. (From Guan, Y.C., Zhou, W., Zheng, H.Y., Lim, G.C., *Appl. Phys. A*, 115, 13–18, 2014. With permission.)

are in agreement with Shimotsuma's work. They fabricated self-organized grating structures, which were in vertical alignment to the polarization direction of incident laser, with the period of 140–320 nm inside silica glass after femtosecond laser irradiation. Moreover, they interpreted this phenomenon in the light of interference between the incident laser light field and the bulk electron plasma wave electric field, causing the periodic modulation of electron plasma concentration and the structure changes of glass.

On the basis of previous studies, the uneven free electron density, caused by surface roughness, might play an important part in periodic structure generation (Dufft et al., 2009, Hsu et al., 2008, Miyaji and Miyazaki, 2006, Wu et al., 2003). Therefore, we propose that the formation of grooves depends on the synthetic functions of original surface roughness and interference between laser and SP. As shown in Figure 5.12 and Figure 5.13, grooves deepened and widened along with the increasing laser pulse number, which is due to the laser fields' transfer to SP

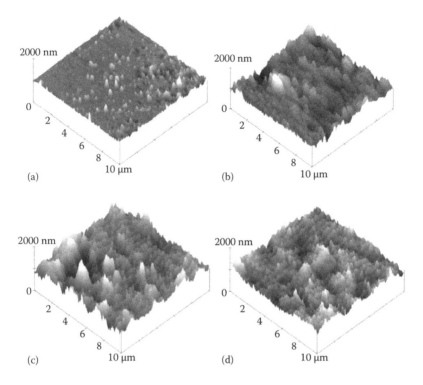

Figure 5.12 AFM images showing the evolution of surface topography on Mg. The probed areas for all the measurements were about 10×10 μm^2. The particulars of surface roughness were depicted in the text: (a) 50 pulses, (b) 200 pulses, (c) 300 pulses, and (d) 500 pulses. (From Guan, Y.C., Zhou, W., Zheng, H.Y., Lim, G.C., *Appl. Phys. A*, 115, 13–18, 2014. With permission.)

(Huang et al., 2009). Moreover, the surface of pure Mg was thermal treated and melted because of its good thermal conductivity (418 W/mK at 20°C) (Avedesian and Baker, 1999, Stull et al., 1955). Surface curling occurred due to the surface tension of the liquid and the severe strain fields caused by the combined effect of highly localized heating and large thermal gradient. Finally, as illustrated in Figure 5.11d and Figure 5.12d, a separate melt (cluster) was formed on the surface.

In summary, the iridescent effect on AZ31B surface irradiated via femtosecond laser is attributed to the structural colors caused by periodic ripples functioning as diffraction gratings. The interference between the incident laser light field and the SP electric field is believed to be the dominant mechanism for the formation of ripples. It is concluded that the current technique may be applicable for other types of Mg alloys by adapting the parameters of laser to their thermal conductivity.

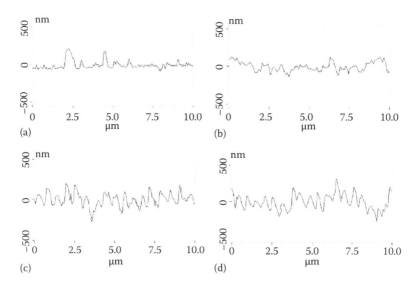

Figure 5.13 AFM images of surface profile measurement in Figure 5.11: (a) 50 pulses, (b) 200 pulses, (c) 300 pulses, and (d) 500 pulses. (From Guan, Y.C., Zhou, W., Zheng, H.Y., Lim, G.C., *Appl. Phys. A*, 115, 13–18, 2014. With permission.)

5.2.3   LMD of Mg alloys

At present, investigation in additive manufacturing technology mainly focuses on titanium, aluminum, and steel alloys. Little public literature is available on laser surface melting of Mg alloys for material build-up (Wei et al., 2014). Mg-rare-earth (RE) alloys where RE elements have the most strengthening effect on magnesium alloys (Fu et al., 2014) are even less studied. In this section, a continuous wavelength CO_2 LMD process is presented to produce a newly developed GW103K Mg-RE alloy (Mg-10Gd-3Y-0.4Zr).

Figure 5.14a shows the gas-atomized powders used for fabrication of LMDed GW103K alloy (Liao et al., 2017). These powders are composed by fine α-Mg grains (6 µm), network-like compounds, and tiny white particles among compounds. Figure 5.14b shows that the as-produced LMDed GW103K alloy consists of the equiaxed grains (19 µm) with clear grain boundaries and the uniformly distributed secondary particles. Such secondary particles mainly distribute at the grain interior. After T6 treatment (solution plus aging treatments), α-Mg grains (158 µm) coarsened greatly and secondary particles are more clearly observed at back-scattered electron mode of SEM, as shown in Figure 5.14c. Energy-dispersive spectroscopy (EDS) results suggest that the particles in Figure 5.14c, especially those with a larger size (Position A), contain a higher amount of Gd and Y than the α-Mg matrix does (Position C), as indicated in Table 5.1 (Liao et al., 2017).

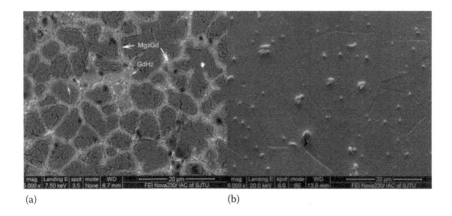

(a)   (b)

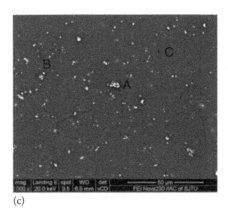

(c)

Figure 5.14 SEM images of gas-atomized GW103K powders (a), as-produced LMDed alloy (b), and T6-treated LMDed alloy at back-scattered electron mode (c). (From Liao, H., Fu, P., Peng, L., Li, J., Zhang, S., Hu, G., Ding, W., *Mater. Sci. Eng. A*, 687, 281–287, 2017. With permission.)

Table 5.1 EDS Results of secondary particles and α-Mg matrix in Figure 3.14b

Position	Mg (wt%)	Gd (wt%)	Y (wt%)
A	85.14	13.33	1.53
B	52.67	25.20	22.12
C	89.91	8.48	1.61

Source: Liao, H., Fu, P., Peng, L., Li, J., Zhang, S., Hu, G., Ding, W., *Mater. Sci. Eng. A*, 687, 281–287, 2017.

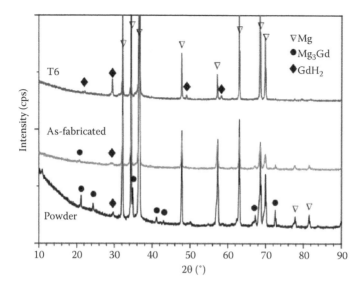

Figure 5.15 XRD results of gas-atomized GW103K powders, as-produced and T6-treated alloys. (From Liao, H., Fu, P., Peng, L., Li, J., Zhang, S., Hu, G., Ding, W., *Mater. Sci. Eng. A*, 687, 281–287, 2017. With permission.)

The X-ray diffraction (XRD) results in Figure 5.15 reveal that after T6 treatment, additional peaks of GdH_2 phase are generated besides that of α-Mg, while little GdH_2 phase is detected in powders and as-produced GW103K alloy (Liao et al., 2017). It is inferred that most of the secondary particles presented in Figure 5.14c are GdH_2 phase, which have also been observed in previous cast Mg-Gd-based alloys (He, 2007, Jiang et al., 2014, Peng et al., 2011). An Mg_3Gd phase is detected in both powders and as-produced alloy; especially, the peaks of Mg_3Gd in powders are much stronger than that in as-produced alloy. Hence, most of the network-like secondary compounds in gas-atomized powders (Figure 5.14a) should be Mg_3Gd phase, and tiny white particles among compounds should be GdH_2 phase. Secondary particles in as-produced alloy (Figure 5.14b) are composed of Mg_3Gd and GdH_2 phases.

Table 5.2 shows the hardness values of LMDed alloy before and after the heat treatment. After T4 treatment (only the solution treatment), hardness decreases from 73.74 to 65.56 HV (Liao et al., 2017). Aging treatment

Table 5.2 Hardness of as-produced and T4- and T6-treated GW103K alloys

State	As-fabricated	T4	T6	ΔHv (T6–T4)
Hardness (Hv)	73.7	65.6	100.5	34.9

Source: Liao, H., Fu, P., Peng, L., Li, J., Zhang, S., Hu, G., Ding, W., *Mater. Sci. Eng. A*, 687, 281–287, 2017.

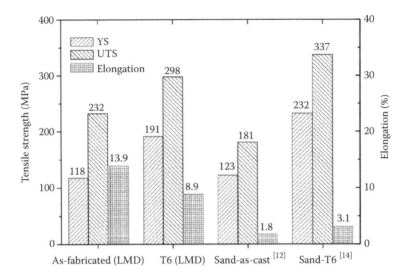

Figure 5.16 Comparison of tensile properties between LMDed and sand cast GW103K alloys before and after T6 treatment. (From Liao, H., Fu, P., Peng, L., Li, J., Zhang, S., Hu, G., Ding, W., *Mater. Sci. Eng. A*, 687, 281–287, 2017. With permission.)

makes hardness further increase to 100.5 HV. The ΔHV between T6- and T4-treated alloys (34.9 HV) is slightly less than that of low-pressure sand cast alloy (38.5 HV). Figure 5.16 shows a comparison of tensile properties between LMDed alloy and previous sand cast alloys (Liao et al., 2017, Jiang et al., 2014, Pang et al., 2013). Compared to as-cast alloys, as-produced LMDed alloy shows higher UTS (+51 MPa) and elongation (+12.1%), as well as slightly lower YS (−5 MPa). T6 treatment makes the YS and UTS of LMDed alloy respectively increase to 191 MPa (+73 MPa) and 298 MPa (+66 MPa), as well as elongation decrease to 8.9% (−5.0%). T6 treatment results in less enhancement of the YS and UTS of LMDed alloy than that of sand cast alloy. Although elongation of LMDed alloy is decreased from 13.9% to 8.9% after T6 treatment, it is still much better than that of sand cast alloys before (1.8%) and after (3.1%) T6 treatment.

As-produced LMDed GW103K alloy shows significantly different microstructure from low-pressure and cast alloys and has finer grains (19 μm) than as-cast alloy (59 μm) (Pang et al., 2013). Secondary phases in as-produced LMDed alloy consist mainly of Mg_3Gd and GdH_2, which are actually hard to be distinguished from each other (Figure 5.14b). In as-cast alloys, the main secondary phase is the Mg_5Gd compound, which presents as network-like morphology and distributes along grain boundaries (Figure 5.17a). These network-like Mg_5Gd compounds have a much larger size than Mg_3Gd particles in as-produced LMDed alloy.

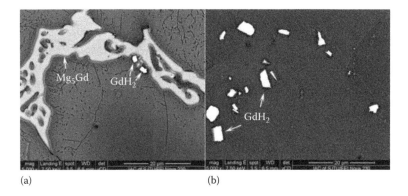

(a)   (b)

Figure 5.17 SEM images of low pressure sand cast GW103K alloy before (a) and after (b) T6 treatment. (From Liao, H., Fu, P., Peng, L., Li, J., Zhang, S., Hu, G., Ding, W., *Mater. Sci. Eng. A*, 687, 281–287, 2017. With permission.)

The dotted-line-like morphologies near network-like Mg_5Gd phase, as shown in Figure 5.17a (Liao et al., 2017), may also be the Mg_5Gd phase precipitated in the casting process due to the low cooling rate. Moreover, large GdH_2 particles are observed at grain boundaries (Figure 5.17a).

After T6 treatment, grains in LMDed alloy grow significantly from 19 to 158 μm, while that in sand cast alloy change from 59 to 88 μm (Jiang et al., 2014). Hence, it is necessary to develop an appropriate heat treatment process for LMDed GW103K alloy to avoid the occurrence of such grain coarsening. After T6 treatment, more XRD peaks of GdH_2 phase have been detected in LMDed alloy than in as-produced alloy (Figure 5.15), suggesting that the amount of GdH_2 particles increases greatly in T6-treated LMDed alloy (Figure 5.14 and Table 5.3), while in sand cast alloy, some large GdH_2 bulks are observed after T6 treatment, as shown in Figure 5.17b (Jiang et al., 2014). Another difference of LMDed and sand cast alloys is the solution content in the matrix, as shown by EDS results in Table 5.3 (Liao et al. 2017).

It has been shown in Figure 5.16 that as-produced LMDed GW103K alloy shows obviously better tensile properties than as-cast alloy does, although the two alloys have comparable YSs. Enhancement of UTS (+51 MPa) and elongation (12.1%) in as-produced alloy is attributed to the refined grains and secondary phases. However, after T6 treatment, LMDed alloy shows obviously lower YS and UTS than sand cast alloy does, although it still has much higher elongation, as indicated in Figure 5.16. Such changes in tensile properties can be explained from three aspects. First, magnesium-vapor-induced serious steam appears during the LMD process due to the very high pressure of Mg vapor. This results in the loss of all the elements in alloy, and the loss rate of Mg is lower than in other

Table 5.3 Microstructure comparison of LMDed and as-cast GW103K alloys before and after T6 treatment

Alloy condition	Grain size (um)	Secondary phases		Solution content in Mg matrix
		Mg_5Gd/Mg_3Gd	GdH_2	
As-fabricated LMDed alloy	19	Mg_3Gd	Small amount, fine particles	4.08% Gd, 0.76% Y
T6 treated LMDed alloy	158	–	Large amount, fine particles	8.48% Gd, 1.61% Y
As-cast sand cast alloy	59	Mg_5Gd	Small amount, large bulk particles	8.79% Gd, 3.58% Y
T6-treated sand cast alloy	88	–	Small amount, large bulk particles	11.98% Gd, 4.04% Y

Source: Liao, H., Fu, P., Peng, L., Li, J., Zhang, S., Hu, G., Ding, W., *Mater. Sci. Eng. A*, 687, 281–287, 2017.

alloying elements. As a result, the content of Gd and Y decreases in as-produced LMDed alloy. Second, grains in LMDed alloy grow significantly from 19 to 158 μm during T6 treatment. Such grain coarsening can result in the reduction of both strength and elongation. Third, the formation of a large number of GdH_2 particles during T6 treatment also consume some Gd and Y atoms, which also reduces Gd and Y content in LMDed alloy. The reduction of Gd and Y content in LMDed alloy results in less precipitates and smaller strengthening effect on YS. The less enhancement of YS in LMDed alloy demonstrates the analysis above. After T6 treatment, the enhancement of YS in LMDed alloy is 73 MPa, which is obviously less than that of sand cast alloy (99 MPa). However, the decreased strengthening effect is beneficial for ductility, which is regarded as the main reason for the higher elongation of T6-treated LMDed alloy (8.9%) than sand cast alloy (3.1%).

In summary, the microstructure and tensile properties of the laser-treated (LMDed) GW103K alloy are affected by the T6 treatment. During the LMD process, a considerable amount of steam occurs due to the very high Mg vapor pressure. The loss rate of Mg is found to be lower than that of the alloying elements. LMD-produced GW103K alloy consists of equiaxed grains (19 μm) and small Mg_3Gd and GdH_2 particles. After the T6 treatment, significant grain coarsening occurs in LMDed alloy with the grain size increased to 158 μm. Small GdH_2 particles are also formed. Compared to the as-cast alloy, as-produced LMDed alloy has slightly lower YS (−5 MPa) and much better UTS (+51 MPa) and elongation (+12.1%).

5.3   Conclusion

The main points presented in this chapter can be summarized as follows.

First, the kinetics of rapid solidification and phase transformation during laser irradiation on AZ91D Mg alloy are described with in-depth analysis. The formation of craters at the irradiated surface is attributed to the laser-induced explosive volume boiling effect. During laser irradiation, the volume expulsion induced melted matter can act as "fresh" defects and cause a next preferential laser irradiation for crater formation. Moreover, alloying elements in Mg alloys are shown to significantly influence the crater formation during the laser processing.

Second, an iridescent effect is generated on the AZ31B surface during femtosecond laser interactions with the Mg alloys. The iridescent effect is attributed to the structural colors caused by periodic ripples functioning as diffraction gratings. The laser-induced periodical ripples are believed to be mainly the interference between the incident laser beam field and the SP electric field.

Finally, during the LMD process, a considerable amount of steam occurs due to the high Mg vapor pressure, which results in the loss of elements in the alloys. The LMDed GW103K alloy consists of equiaxed grains with a diameter of 19 μm and small Mg_3Gd/GdH_2 particles. Compared with the low-pressure sand-cast GW103K alloy, the LMDed alloy shows slightly lower YS (−5 MPa) but better UTS (+51 MPa) and elongation (+12.1%). T6 treatment results in significant grain coarsening in the LMDed GW103K alloy, as well as noticeable enhancement of YS and UTS.

Acknowledgments

The authors would like to thank the National Natural Science Foundation of China (grant number 51705013), the National Program of Key Research in Additive Manufacturing and Laser Manufacturing of China (grant number 2016YFB1102503), National Key Research and Development Program of China (2016YFB0301000 and 2016YFB0701204), Beijing Natural Science Foundation (grant number 3162019), Agency for Science, Technology and Research (A*STAR) Remanufacturing Programme (project code U11-M-030AU&U09-M-006SU) A*STAR, Singapore, Remanufacturing Programme (SERC grant no: 112 290 4017), and Nanyang Technological University (PhD Scholarship) for support of the current work.

References

Avedesian, M. M., Baker, H. 1999. *ASM specialty handbook: Magnesium and magnesium alloys*, ASM International, Materials Park, OH.

Bhattacharya, D., Singh, R., Holloway, P. 1991. Laser–target interactions during pulsed laser deposition of superconducting thin films. *Journal of Applied Physics*, 70, 5433–5439.

Bonse, J., Sturm, H., Schmidt, D., Kautek, W. 2000. Chemical, morphological and accumulation phenomena in ultrashort-pulse laser ablation of TiN in air. *Applied Physics A: Materials Science & Processing*, 71, 657–665.

Bonse, J., Munz, M., Sturm, H. 2005. Structure formation on the surface of indium phosphide irradiated by femtosecond laser pulses. *Journal of Applied Physics*, 97, 013538.

Carou, D., Rubio, E. M., Lauro, C. H., Davim, J. P. 2014. Experimental investigation on surface finish during intermittent turning of UNS M11917 magnesium alloy under dry and near dry machining conditions. *Measurement*, 56, 136–154.

Chen, Y., Samant, A., Balani, K., Dahotre, N. B., Agarwal, A. 2009. Effect of laser melting on plasma-sprayed aluminum oxide coatings reinforced with carbon nanotubes. *Applied Physics A*, 94, 861.

Craciun, V., Bassim, N., Singh, R., Craciun, D., Hermann, J., Boulmer-Leborgne, C. 2002. Laser-induced explosive boiling during nanosecond laser ablation of silicon. *Applied Surface Science*, 186, 288–292.

Dou, K., Parkhill, R. L., Wu, J., Knoobe, E. 2001. Surface microstructuring of aluminum alloy 2024 using femtosecond excimer laser irradiation. *IEEE Journal of Selected Topics in Quantum Electronics*, 7, 567–578.

Dufft, D., Rosenfeld, A., Das, S., Grunwald, R., Bonse, J. 2009. Femtosecond laser-induced periodic surface structures revisited: A comparative study on ZnO. *Journal of Applied Physics*, 105, 034908.

Fu, P., Peng, L., Jiang, H., Ding, W., Zhai, C. 2014. Tensile properties of high strength cast Mg alloys at room temperature: A review. *China Foundry*, 11, 277–286.

Guan, Y., Zhou, W., Li, Z., Zheng, H. 2009a. Study on the solidification microstructure in AZ91D Mg alloy after laser surface melting. *Applied Surface Science*, 255, 8235–8238.

Guan, Y., Zhou, W., Zheng, H. 2009b. Effect of laser surface melting on corrosion behaviour of AZ91D Mg alloy in simulated-modified body fluid. *Journal of Applied Electrochemistry*, 39, 1457–1464.

Guan, Y., Zhou, W., Zheng, H. 2009c. Effect of Nd: YAG laser melting on surface energy of AZ91D Mg alloy. *Surface Review and Letters*, 16, 801–806.

Guan, Y., Zhou, W., Zheng, H., Li, Z. 2010a. Solidification microstructure of AZ91D Mg alloy after laser surface melting. *Applied Physics A: Materials Science & Processing*, 101, 339–344.

Guan, Y. C., Zhou, W., Zheng, H. Y., Li, Z. L. 2010b. Surface modification of AZ91D magnesium alloy using millisecond, nanosecond and femtosecond lasers. key engineering materials. *Key Engineering Materials*, 447–448, 695–699.

Guan, Y. C., Zhou, W., Li, Z. L., Zheng, H. Y. 2013. Femtosecond laser-induced iridescent effect on AZ31B magnesium alloy surface. *Journal of Physics D: Applied Physics*, 46, 425305.

Guan, Y. C., Zhou, W., Zheng, H. Y., Lim, G. C. 2014a. Femtosecond laser-induced ripple structures on magnesium. *Applied Physics A*, 115, 13–18.

Guan, Y. C., Zhou, W., Li, Z. L., Zheng, H. Y. 2014b. Boiling effect in crater development on magnesium surface induced by laser melting. *Surface Coatings Technology*, 252, 168–172.

He, S. M. 2007. Study on the microstructural evolution, properties and fracture behavior of Mg-Gd-Y-Zr(-Ca) Alloys. PhD thesis, Shanghai Jiao Tong
University.

Hou, J., Zhou, W., Zhao, N. 2010. Methods for prevention of ignition during
machining of magnesium alloys. *Advanced Precision Engineering*, 447–448,
150–154.

Hsu, E., Crawford, T., Maunders, C., Botton, G., Haugen, H. 2008. Cross-sectional
study of periodic surface structures on gallium phosphide induced by ultra-
short laser pulse irradiation. *Applied Physics Letters*, 92, 221112.

Huang, M., Zhao, F., Cheng, Y., Xu, N., Xu, Z. 2009. Origin of laser-induced near-
subwavelength ripples: Interference between surface plasmons and incident
laser. *ACS Nano*, 3, 4062–4070.

Itina, T. E., Gouriet, K., Zhigilei, L. V., Noël, S., Hermann, J., Sentis, M. 2007.
Mechanisms of small clusters production by short and ultra-short laser ablation. *Applied Surface Science*, 253, 7656–7661.

Jacobs, J. W., Rikken, J. M. 1987. Boiling Effects and bubble formation at the
solid–liquid interface during laser-induced metal deposition. *Journal of the
Electrochemical Society*, 134, 2690–2696.

Jiang, L., Liu, W., Wu, G., Ding, W. 2014. Effect of chemical composition on the
microstructure, tensile properties and fatigue behavior of sand-cast Mg-Gd-
Y-Zr alloy. *Materials Science and Engineering: A*, 612, 293–301.

Kannan, M. B., Raman, R. S. 2008. In vitro degradation and mechanical integrity
of calcium-containing magnesium alloys in modified-simulated body fluid.
Biomaterials, 29, 2306–2314.

Li, Z., Zheng, H., Teh, K., Liu, Y., Lim, G., Seng, H., Yakovlev, N. 2009. Analysis of
oxide formation induced by UV laser coloration of stainless steel. *Applied
Surface Science*, 256, 1582–1588.

Liao, H., Fu, P., Peng, L., Li, J., Zhang, S., Hu, G., Ding, W. 2017. Microstructure
and mechanical properties of laser melting deposited GW103K Mg-RE alloy.
Materials Science & Engineering A, 687, 281–287.

Miyaji, G., Miyazaki, K. 2006. Ultrafast dynamics of periodic nanostructure formation on diamondlike carbon films irradiated with femtosecond laser
pulses. *Applied Physics Letters*, 89, 191902.

Pang, S., Wu, G., Liu, W., Sun, M., Zhang, Y., Liu, Z., Ding, W. 2013. Effect of cooling rate on the microstructure and mechanical properties of sand-casting
Mg-10Gd-3Y-0.5Zr magnesium alloy. *Materials Science and Engineering: A*,
562, 152–160.

Peercy, P. 1986. Solidification dynamics and microstructure of metals in pulsed
laser irradiation. *Laser Surface Treatment of Metals*, 115, 57–78.

Peng, Q., Huang, Y., Meng, J., Li, Y., Kainer, K. U. 2011. Strain induced GdH_2 pre-
cipitate in Mg-Gd based alloys. *Intermetallics*, 19, 382–389.

Pereira, A., Cros, A., Delaporte, P., Georgiou, S., Manousaki, A., Marine, W., Sentis,
M. 2004. Surface nanostructuring of metals by laser irradiation: Effects of
pulse duration, wavelength and gas atmosphere. *Applied Physics A: Materials
Science & Processing*, 79, 1433–1437.

Rubahn, H.-G. 1999. *Laser applications in surface science and technology*, John Wiley
& Sons.

Schmidt, V., Husinsky, W., Betz, G. 2000. Dynamics of laser desorption and ablation of metals at the threshold on the femtosecond time scale. *Physical Review
Letters*, 85, 3516.

Steen, W. 2003. Laser material processing—An overview. *Journal of Optics A: Pure and Applied Optics*, 5, S3.

Stull, D. R., Mcdonald, R. A. 1955. The enthalpy and heat capacity of magnesium and of type 430 stainless steel from 700 to 1100°K. *Journal of the American Chemical Society*, 77, 5293–5293.

Tillack, M., Blair, D., Harilal, S. 2004. The effect of ionization on cluster formation in laser ablation plumes. *Nanotechnology*, 15, 390.

Vorobyev, A., Guo, C. 2008. Spectral and polarization responses of femtosecond laser-induced periodic surface structures on metals. *Journal of Applied Physics*, 103, 043513.

Wei, K., Gao, M., Wang, Z., Zeng, X. 2014. Effect of energy input on formability, microstructure and mechanical properties of selective laser melted AZ91D magnesium alloy. *Materials Science and Engineering: A*, 611, 212–222.

Wu, Q., Ma, Y., Fang, R., Liao, Y., Yu, Q., Chen, X., Wang, K. 2003. Femtosecond laser-induced periodic surface structure on diamond film. *Applied Physics Letters*, 82, 1703–1705.

Yang, Z., Debroy, T. 1999. Modeling macro-and microstructures of gas-metal-arc welded HSLA-100 steel. *Metallurgical and Materials Transactions B*, 30, 483–493.

Zhang, X., Chu, S., Ho, J., Grigoropoulos, C. 1997. Excimer laser ablation of thin gold films on a quartz crystal microbalance at various argon background pressures. *Applied Physics A: Materials Science & Processing*, 64, 545–552.

Zheng, H., Qian, H., Zhou, W. 2008. Analyses of surface coloration on TiO 2 film irradiated with excimer laser. *Applied Surface Science*, 254, 2174–2178.

Zhou, W., Long, T., Mark, C. 2007. Hot cracking in tungsten inert gas welding of magnesium alloy AZ91D. *Materials Science and Technology*, 23, 1294–1299.

chapter six

Sensor monitoring of titanium alloy machining

Roberto Teti and Alessandra Caggiano

Contents

6.1 Introduction

Titanium alloys exhibit excellent properties, such as a high strength-to-weight ratio, medium density, exceptional resistance to corrosion, low coefficient of thermal expansion, and high toughness, that make them particularly suitable for advanced applications such as those in the aerospace, automotive, and medical industries (Ezugwu et al. 2003; M'Saoubi et al. 2015).

In the aerospace sector, the ability of these alloys to maintain their high strength at high operating temperatures makes them qualified materials for aircraft engine components, including low- and high-pressure compressors, disks, and blades. Moreover, these alloys are suitable for airframe structures where the operating temperature exceeds 130°C (which is the conventional maximum operating temperature for aluminium alloys). The exceptional resistance to corrosion provides also savings on the employment of protective coatings like paints (Ezugwu et al. 2003). In modern aircrafts such as the Boeing B787 and Airbus A350XWB, novel titanium alloys are extensively employed because they are characterized by elevated compatibility with carbon fiber reinforced composite materials (M'Saoubi et al. 2015).

In the last years, new titanium alloys have been developed to achieve weight savings in diverse applications, especially in the aerospace sector. The most widely employed Ti alloys are the alpha (α), beta (β), and alpha/beta (α/β) Ti-6Al-4V. The β tempered Ti-6Al-4V alloy, with high performance in terms of damage tolerance, is generally selected for a number of applications such as underwing fittings, landing gear beams, centre wing box structures, etc. Ti-3Al-2.5V titanium sheets, which display elevated formability, are employed for producing anticrash structure components. The Ti-10V-2Fe-3Al alloy has been utilized to replace the high-strength 4340M alloy steel in landing gears of the Boeing 777 aircraft, while Ti-5Al-5V-5Mo-3Cr (Ti-5553), which is the new-generation high-strength β titanium alloy, has been utilized in the Boeing 787, as it represents a promising material for advanced structural and landing gear applications, with higher mechanical properties compared with the traditional Ti-6Al-4V alloys (Arrazola et al. 2009; M'Saoubi et al. 2015).

These advanced high-temperature alloys require high-performance machining solutions and methods to satisfy the requirements in terms of accuracy, complex free form geometries, and surface integrity, which are notably challenging taking into account the low machinability of these alloys (M'Saoubi et al. 2015).

6.2   Machinability of titanium alloys

Machinability is defined as the ease (or difficulty) with which a material can be machined under a given set of operating conditions, including cutting speed, feed rate, and depth of cut. The machinability of a material is mainly assessed based on tool life, surface finish, and power consumed during the machining operation, but also component forces and chip shape are useful indicators (Ezugwu et al. 2003).

The machinability of titanium and its alloys is generally considered to be poor owing to several inherent properties of the materials.

Titanium is very chemically reactive and therefore has a tendency to weld to the cutting tool during machining, thus leading to chipping and premature tool failure. Its low thermal conductivity, which is about 1/6 that of steels, increases the temperature at the tool–workpiece interface up to about 1100°C, which affects the tool life adversely. As a matter of fact, a large proportion (about 80%) of the heat generated when machining titanium alloys is conducted into the tool because it cannot be bed into the workpiece due to the low thermal conductivity. Additionally, its high strength maintained at elevated temperature and its low modulus of elasticity further impair its machinability (Ezugwu and Wang 1997). The low modulus of elasticity, which rapidly drops even under moderate temperatures, is responsible for workpiece bending, particularly when machining components with thin walls (Lopez de Lacalle et al. 2000). The chip formation process is unstable and typically produces a segmented, or cyclical, chip. Chatter phenomena often occur due to the variations of the cutting force in both the axial and cutting directions; therefore, a very stiff tool-workpiece-machine system is required.

With regard to the cutting speed, which affects temperature and tool wear, titanium alloys are generally hard to machine under cutting speeds >30 m/min using high-speed steel tools, and under cutting speeds >60 m/min using cemented tungsten carbide tools. Due to the high reactivity with titanium alloys, other categories of materials, such as ceramic, cubic boron nitride, and diamond, cannot be employed as tool materials for machining of these alloys (Lopez de Lacalle et al. 2000).

Because of the previously mentioned critical issues arising during machining, a number of efforts have been spent to optimize the machining processes of the low-machinability titanium alloys. In this framework, the development of advanced sensor monitoring procedures can significantly improve the machining processes. In the last decades, the use of sensor systems for online monitoring of tool conditions, chip formation, surface integrity, process conditions, and chatter detection has been proposed. In the following section, the major sensor monitoring applications in machining of titanium alloys are presented.

6.3   Sensor monitoring of titanium alloy machining

When dealing with low-machinability materials such as titanium alloys, the development of sensor monitoring procedures can significantly improve the process performance in terms of part quality and tool wear, as well as increase productivity and reduce production costs.

To improve and optimize machining processes, a number of studies in the last decades proposed the use of sensor systems for online monitoring of tool conditions, machine tool state, chip formation, surface integrity, process conditions, chatter detection, etc. (Teti et al. 2010).

For instance, TCM during machining allows for a reduction of tool costs, optimizing tool life by implementing condition-based tool replacement strategies (i.e., by replacing tools only when they are close to end of life) instead of conservative time-based tool replacement strategies (in which the tool is replaced after a predetermined time independently of its real wear conditions), and helps reduce machine and workpiece damage risk by allowing fast reaction when a tool breakage occurs (Caggiano et al. 2016).

Machining process monitoring systems work on the basis of a fundamental underlying principle. During material cutting, there is a number of process variables, including vibrations, cutting force, acoustic emission (AE), temperature, etc., which are affected by cutting tool and machining process conditions. Through the employment of suitable physical sensors, it is possible to measure those variables that are potentially effective for monitoring the machining process. The sensor signals acquired by such physical sensors are then conditioned and processed with the aim to extract a number of useful sensor signal features, which are correlated with tool state and process conditions (Teti et al. 2010).

To exploit the sensor monitoring results as a support to decision-making systems for machining processes, the development of advanced sensor monitoring procedures, based on innovative technologies and approaches, is essential. The sensor signal features extracted from the acquired sensor signals are then sent as input to cognitive decision-making support systems, which elaborate the features to carry out the final diagnosis. The latter can be transmitted to the machine tool operator or directly sent to the computer numerical control (CNC) of the machine tool in order to suggest or implement the suitable adaptive or corrective actions (Teti et al. 2010).

6.3.1   Sensors and sensor monitoring systems for machining process monitoring

A number of sensors and sensor monitoring systems have been developed in the last years for machining process monitoring. The main requirements can be summarized as follows: measurement position as close as possible to the cutting zone; no constraint of working volume and cutting parameters; no decrease in the static and dynamic machine tool stiffness; low wear and maintenance, easy replaceability, and low cost; resistance to dirt and chips, as well as mechanical, electromagnetic and thermal stresses, etc.

   In the followings, the major machining process monitoring sensors are presented according to a classification based on the measured variable.

6.3.1.1   Force and torque sensor

The monitoring of cutting forces, responsible for material removal in conventional machining processes, has been extensively employed by researchers for several scopes, such as the validation of analytical models of the process, monitoring of tool condition, etc., due to the high sensitivity and quick response of the force signals to variations concerning the cutting condition.

   Force sensors consist of a mechanical structure that responds to an applied force, while torque sensors respond to a torsional load. Both types of sensors typically employ sensing units that translate the applied force or torsional load into an elastic element deformation. The most commonly used force and torque sensors are piezoelectric and strain-based sensors (Byrne et al. 2004; Teti et al. 2010).

6.3.1.2   Vibration sensor

Vibrations occurring in machining processes can be classified into dependant and independent of the cutting process. Independent vibrations comprise forced vibration due to other machines or machine components, such as vibration transmitted via foundations, inertia forces in reciprocating parts, rotating unbalance, and lack of kinematic accuracy of drives. On the other hand, vibrations dependent on the cutting process are related to tool engagement conditions during machining, which significantly affect the generated vibration. The most renowned type of vibration in machining is chatter, which happens due to the waviness regeneration caused by the interaction between material surface and tool at specific spindle rotational frequencies and has a negative impact on surface finish and tool life. Although a wide variety of sensing principles are available to sense vibration, piezoelectric transducers are the most common vibration sensors employed in monitoring of machining operations (Teti et al. 2010; Segreto et al. 2017).

6.3.1.3   AE sensor

AE sensors with wide sensor dynamic bandwidth in the range of 100–900 kHz are able to perceive most of the phenomena in cutting processes. Data acquisition and signal processing require substantial efforts related to signal processing and the use of bandpass filters with appropriate frequency ranges. The signal obtained from the AE sensor is generally preamplified and processed by a root mean square (RMS) converter, and suitable gains and filters need to be set. The installation of AE sensors requires the application of a couplant between the sensor and the material surface, which should be clean and without paint or other barriers that could affect the acoustic coupling. As the signal attenuation becomes higher when the sensor is placed far from the AE source, in the case that the AE sensor is installed on the workpiece, the variation in the sensor–source distance during the machining process needs to be taken into account. The most commonly employed AE sensors for machining process monitoring are piezoelectric sensors (Teti et al. 2010).

6.3.1.4   Power sensor

The mechanical force required to perform material removal is provided by electric drives and spindles. Hence, the measurement of motor power or current allows monitoring both the process power as well as the machine tool and drive conditions. In the case of machining process monitoring based on the measurement of motor power or current, the sensor system does not disturb the process itself as the capability to measure power is already available in the drive controller or can be easily retrofitted and can be easily employed in the industrial production context (Byrne et al. 1995; Teti et al. 2010).

6.3.1.5   Other sensor types

A number of other sensor types have been employed in the literature, including for instance temperature sensors or vision systems. The use of temperature measurement in machining was comprehensively reviewed by Davies et al. (2007), while the employment of vision systems for monitoring tool condition was systematically reported by Kurada and Bradley (1997).

6.3.2   Sensor monitoring applications in the machining processes of titanium alloys

To improve and optimize the machining processes applied to low-machinability titanium alloys, a number of studies in the last decades proposed the use of sensor systems for online monitoring of tool conditions, chip formation, surface integrity, process conditions, chatter detection, etc.

In this section, the main studies dealing with sensor monitoring applied to the major machining processes of titanium alloys are reported. For each category of machining process, the applications are classified based on the sensor monitoring scope.

6.3.2.1   Milling

6.3.2.1.1   Cutting force-vibration monitoring   Machining processes like milling, distinguished by interrupted cutting, are typically vulnerable to issues related to the vibration of the machine tool–workpiece fixture systems because the frequency of tool entry on the workpiece is often close to their natural frequency harmonics. This phenomenon is considerably accentuated when milling titanium alloys, since the latter display a low Young modulus and, consequently, an extended elastic behavior, which causes notable variations in chip thickness and instable cutting forces. Moreover, the extremely low heat conductivity is responsible for the development of serrated chips, which further amplify the instability of the cutting forces.

In Ítalo Sette Antonialli et al. (2010), vibration analysis based on cutting force was applied to verify the influence of the tool entering angle on the tool vibration and, therefore, on tool life in milling of titanium alloys. Through the tool vibration analysis tests, it was possible to identify the frequency range in which the tool vibration is amplified; that for the specific milling operation was >400 Hz. The radial component of the cutting force was processed and analyzed in the frequency domain for each experimental milling test, showing an amplified load rate within the 400–1000 Hz high-frequency range. For all the experimental testing conditions, the tool life and wear development were examined and correlated to the tool vibration in that frequency range. It was shown that the employment of a tool with a greater entering angle and round inserts relates the radial load to higher frequencies, at which the tool does not behave in a rigid manner, causing tool life shortening due to cutting edge breakage. On the other hand, a smaller entering angle produced normal wear mechanisms based on diffusion and attrition (Ítalo Sette Antonialli et al. 2010).

6.3.2.1.2   Chatter detection   Chatter is especially severe when finishing titanium alloys, due to their low Young's modulus and the elastic behavior of the workpiece material, making it spring from the cutting tool, rub the cutting edges together, hence intensifying friction and further increasing the temperature in the cutting zone as well as the vibrations. The segmented chips contribute to additionally increase the cutting force fluctuations, generating vibration or chatter in the cutting process.

Chatter may cause fast wear of tools and poor surface quality of the workpieces at high cutting speed, and it will happen under different process parameters. In Huang et al. (2012), a sensor monitoring procedure based on the acquisition and analysis of milling force and acceleration sensor signals was developed to identify chatter, with the aim to select the suitable cutting speed to suppress chatter. Through experimental milling tests, the performance of machining vibration in milling of Ti-6Al-4V alloy using end mill with variable pitch under different cutting speeds in the range of 80–360 m/min was investigated. The milling force and acceleration signals acquired during the milling tests were examined with the aim to compare their behavior under stable and unstable milling process conditions. The obtained results provided information concerning the frequency of chatter and the milling force at different frequencies.

The experimental results showed that in the presence of chatter, the milling forces noticeably increased by 61.9%–66.8% compared to stable cutting. Moreover, the quality of the machined surface worsened and the surface roughness grew by 34.2%–40.5% compared to the surface roughness obtained under stable cutting. Finally, a thorough analysis of the milling force and the machined surface quality allowed identifying the optimal cutting speed for milling Ti-6Al-4V alloy under the given parameter conditions.

Chatter vibrations are a common problem especially in milling of thin-walled workpieces due to the lack of dynamic stiffness. In Feng et al. (2016), the fast Fourier and wavelet transform methods were employed to analyze the cutting force signals in the direction normal to the machined surface with the aim to detect chatter in vertical milling of a thin-walled Ti-6Al-4V alloy workpiece. To study the chatter phenomenon, the topography of the machined surface and the prediction theory of regenerative chatter were jointly employed. The results showed that fast Fourier transform (FFT), wavelet transform, as well as the combination of surface topography and regenerative chatter prediction theory can successfully detect chatter. Moreover, it was possible to identify the stable cutting zone via simulation using a stability lobe diagram, with the aim to help in the selection of the appropriate cutting parameters to avoid chatter.

In Huang et al. (2016), milling tests were performed on Ti-6Al-4V thin-wall components and non-thin-wall components under the same cutting parameters to evaluate the structure impact on cutting force and spindle vibration at different tool wear values by monitoring cutting force and spindle acceleration signals. In order to observe the variation of frequency components and wavelet coefficients with the tool wear development, frequency spectrum and wavelet analysis methods were employed. The experimental results illustrated that, when milling thin-wall components, the cutting vibration was notably higher than in the case of non-thin-wall components due to the weak rigidity, which generated relatively low cutting force for the lower cutting load. In particular, when milling thin-wall components, the a_y acceleration component was higher because of their weak rigidity, while the cutting force F_y decreased because of the decline of cutting load. Compared to the other two directions, the value of the a_y acceleration component and the F_y cutting force component notably increased with tool wear development.

6.3.2.1.3 Temperature monitoring As previously mentioned, critical issues in machining of titanium alloys are related to the elevated temperature, which is concentrated in a limited area at the tool tip, and the formation of segmented chips, which are characterized by adiabatic shear bands, as a result of mechanical instability. The high tool temperature is responsible for accelerated tool wear, while the formation of chip shear band generates cutting force fluctuations and chipping at the tool cutting edges. Moreover, it has been illustrated that the cutting temperature has a direct effect on the surface integrity of the machined workpiece and the machining accuracy. Therefore, the understanding of temperature development in machining of titanium alloys represents an important study to improve cutting process performance (Sun et al. 2014).

In milling, tool wear can be influenced by cutting interruptions that, on the one hand, lower the temperatures at the tool–chip interface but, on the other hand, produce thermal and stress cycling. To determine the temperature during interrupted cutting of titanium alloy Ti-6Al-4V, microthermal imaging was used in Armendia et al. (2010). Machining tests under cutting speeds up to 180 and 640 m/min using TiAlN/TiN-coated carbide milling inserts were performed. The microthermal imaging technique allows spatial mapping of the thermal fluctuations on the tool, which may be crucial to detect the reasons for tool failure. This method allows measuring the localized maximum temperature on the tool with a resolution of 30 μm, instead of the spatially averaged maximum temperatures measurable by the systems previously reported in the literature (tool chip thermocouple and color pyrometer). The map of the thermal cycling experienced by a tool in a single cutting cycle of Ti-6Al-4V alloy at 80 m/min and 50% time-in-cut illustrates that the largest temperature variation is localized on the tool tip, where a thermal cycling about 250°C can be detected. Thermal cycling has a key role in the development of tool wear, coating delamination, and failure in milling due to chipping and cracking of tool cutting edges. This type of thermal cycling map is achievable only with a high-resolution imaging technique such as the one presented by Armendia et al. (2010).

6.3.2.2 Turning

6.3.2.2.1 Tool wear monitoring   As illustrated previously, extremely rapid tool wear and short tool life are experienced in machining of titanium alloys due to the high cutting temperature and the strong adhesion at the tool–chip interface and tool–workpiece interface caused by the low thermal conductivity and high chemical reactivity of these alloys.

With the aim to monitor the tool wear state during dry turning of Ti-6Al-4V alloy, a cognitive sensor monitoring procedure based on the acquisition and processing of cutting force, AE, and vibration signals during turning was implemented in Caggiano et al. (2017).

The developed procedure was based on sensor signal feature extraction and selection and the employment of a cognitive pattern recognition method based on artificial neural networks (ANN), allowing for an accurate diagnosis of tool wear state. This diagnosis can be used for tool replacement strategies based on actual tool wear state instead of preventive strategies, allowing fully exploiting the entire tool life.

The experimental testing campaign consisted of consecutive cylindrical turning passes of 100 mm length under different cutting conditions (cutting speed, v = 60, 70, and 80 m/min; feed rate, f = 0.10, 0.25, and 0.30 mm/rev; depth of cut, d = 0.5, 1.0, and 1.5 mm) on 60-mm-diameter Ti-6Al-4V alloy bars. The CNC lathe was equipped with a multiple sensor system (Figure 6.1) including a Montronix FS1xCXK-x-ICA three-dimensional (3D) force sensor, a Montronix BV100 AE sensor, and a Montronix Spectra Pulse 3D vibration sensor. The first two sensors, providing analog signals, were connected to a NI USB-6361 DAQ board for digitization of the three cutting force components (F_x, F_y, and F_z) and the RMS of AE (AE_{RMS}) signal, while the wireless digital 3D vibration sensor sent directly to the personal computer the sensorial data corresponding to the three vibration acceleration components (A_x, A_y, and A_z). The sampling rate on the DAQ board was set to 10 kS/s.

The cutting tool employed during the experimental turning tests was a Mitsubishi CNMG120404-MS MT9015 turning insert, and no cutting fluid was employed. The final aim of the sensor monitoring procedure was to perform TCM through the identification of correlations between the multiple sensor signals acquired during dry turning of Ti-6Al-4V and the tool wear conditions. After each turning pass, a magnified picture

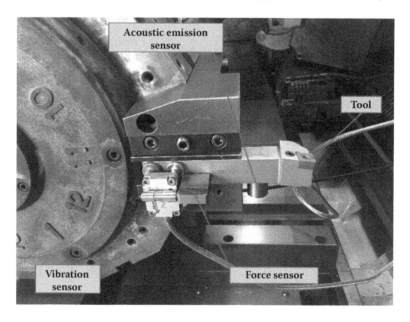

Figure 6.1 Multiple sensor system mounted on the tool holder. (From Caggiano, A., Napolitano, F., Teti, R., *Procedia CIRP*, 62, 209–214, 2017. With permission.)

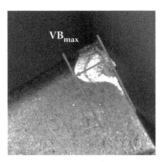

Figure 6.2 Measurement of maximum flank wear land, VB_{max}. Cutting parameters: v = 60 m/min, f = 0.3 mm/rev, d = 1 mm. (From Caggiano, A., Napolitano, F., Teti, R., *Procedia CIRP*, 62, 209–214, 2017. With permission.)

of the cutting edge was acquired through a portable digital microscope (DINO-LITE Premier) to measure the maximum flank wear land, VB_{max}, as shown in Figure 6.2. The criterion for the end of tool life was established by setting a maximum acceptable value VB_{max} = 0.6 mm.

The results on tool wear development confirm that cutting speeds 70 m/min and 80 m/min were too high for dry turning of Ti-6Al-4V. As a matter of fact, by boosting the cutting speed, a large increase in tool wear rate and a notable reduction in machining time and material removal carried out with a single tool were observed. With regard to the differences between tool wear development at 60 m/min and 70 m/min, they were smaller at low feed rate and depth of cut but significantly increased with growing feed rate and depth of cut.

On the basis of the experimental results, 60 m/min appeared as the most suitable value for maximum cutting speed in dry turning of Ti-6Al-4V. By setting the cutting speed at 60 m/min, a notable growth of tool wear rate (3 times faster) was found when increasing depth of cut from d_1 = 0.5 mm to d_3 = 1.5 mm, as well as when increasing feed rate from f_1 = 0.20 mm/rev to f_3 = 0.3 mm/rev.

The signals acquired via the multiple sensor monitoring system were preprocessed and segmented to isolate the relevant signal portion corresponding to actual machining with the aim to allow for the subsequent extraction of functional sensor signal features to be correlated with tool state.

A conventional statistical approach in the time domain was applied for feature extraction, and the following five statistical signal features were taken into consideration: arithmetic mean, variance, skewness, kurtosis, and signal power. For each extracted feature, the Pearson's correlation coefficient, r, was calculated to evaluate the correlation with tool wear conditions, with the aim to select only a limited number of features highly correlated with tool wear. Based on the r value, only eight out of all the extracted statistical features showed a strong correlation with tool wear: F_x average (F_{x_av}),

F_y average (F_{y_av}), F_z average (F_{z_av}), AE_{RMS} average (AE_{RMS_av}), F_x skewness (F_{x_sk}), F_z skewness (F_{z_sk}), F_x kurtosis (F_{x_kurt}), and F_z kurtosis (F_{z_kurt}).

The selected statistical features were employed to construct sensor fusion feature pattern vectors (SFPVs) to be correlated with tool wear state through cognitive pattern recognition based on ANNs. ANN paradigms were developed to accomplish two main objectives: (a) reconstruction of the missing points of a single tool wear curve obtained from several passes under a given turning condition and (b) generation of the entire tool wear curve for a given turning condition based on a training set consisting of sensor signal features and corresponding tool wear values measured for different turning conditions.

In both cases, three-layer cascade-forward backpropagation ANNs using the Levenberg-Marquardt optimization algorithm for training were set up, and an algorithm to train and test different numbers of hidden layer nodes was implemented. The pattern recognition performance of the ANN was assessed in terms of mean squared error (MSE) between the ANN predicted VB_{max} values and the measured VB_{max} values.

In the case of ANN tool wear curve reconstruction, the ANNs were fed with a set of nine-feature $SFPV_a$ constructed for each turning pass by combining the selected statistical signal features and the corresponding machining time (t), i.e., the time at which the turning pass was terminated.

The ANN tool wear curve reconstruction provided very low MSE values between 0.00064 and 0.02275, achieving an accurate tool wear curve reconstruction with ANN output values very close to the measured tool wear values. Figure 6.3 shows the ANN output vs. measured VB_{max} for the turning test at v = 60 m/min, f = 0.25 mm/rev, and d = 0.5 mm.

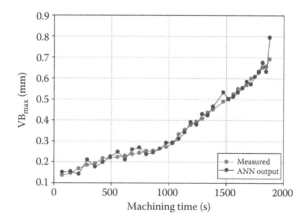

Figure 6.3 ANN output vs. measured VB_{max} for turning test at v = 60 m/min, f = 0.25 mm/rev, d = 0.5 mm. ANN configuration: 9-27-1. MSE = 0.00524. (From Caggiano, A., Napolitano, F., Teti, R., *Procedia CIRP*, 62, 209–214, 2017. With permission.)

With regard to the entire tool wear curve generation, the new $SFPV_b$ was a 12-feature vector constructed by adding to $SFPV_a$ the cutting parameters, that is, cutting speed, feed rate, and depth of cut. In this case, slightly higher MSE values between 0.00613 and 0.02683 were obtained, suggesting that this task is more challenging for the ANN. Figure 6.4 shows the tool wear curve for turning with $v_1 = 60$ m/min, $f_1 = 0.25$ mm/rev, and $d_1 = 1.5$, generated by the ANN trained with signal features and corresponding tool wear values for surrounding turning conditions. Although a higher MSE = 0.01373 was obtained in this case, the ANN was still able to satisfactorily provide the tool wear development and identify the critical moment at which the transition of the tool wear curve between second and third tool wear zone occurred.

The results obtained in Caggiano et al. (2017) confirmed the capability of reliably carrying out a diagnosis on tool wear state through the developed multisensor monitoring methodology for turning of Ti6-Al-4V titanium alloys based on the acquisition of sensor signals related to cutting force, vibration and AE, and cognitive pattern recognition via ANN. The tool wear state diagnosis can be employed to support decision making for appropriate corrective actions on tool replacement, parameter variation, or process stop, which can be either directly sent as input to the CNC of the machine tool or recommended to the human operator.

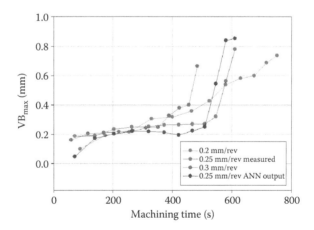

Figure 6.4 ANN entire tool wear curve generation. The tool wear curve for $v_1 = 60$ m/min, $f_1 = 0.25$ mm/rev, $d_1 = 1.5$ mm is obtained using a training set comprising sensor signal features and corresponding tool wear values for $v_2 = 60$ m/min, $f_2 = 0.20$ mm/rev, $d_2 = 1.5$ mm and $v_3 = 60$ m/min, $f_3 = 0.30$ mm/rev, $d_3 = 1.5$ mm. ANN configuration: 12-24-1. MSE = 0.01373. (From Caggiano, A., Napolitano, F., Teti, R., *Procedia CIRP*, 62, 209–214, 2017. With permission.)

6.3.2.2.2   Catastrophic tool failure detection   A catastrophic tool failure (CTF) is an unpredictable fracture of the cutting edge, e.g., through brittle failure, that may produce considerable damages to the workpiece and the machine tool and is particularly critical when machining difficult-to-machine materials such as titanium alloys. The ability for online detection of a CTF event during machining and the prompt halting of the process are essential to reduce workpiece scraps and machine downtime with positive effects on product quality and cost.

In the literature, a number of procedures to identify the occurrence of CTF were proposed with particular reference to turning processes (Jemielniak and Otman 1998a, 1998b; Balsamo et al. 2016). Such methodologies are based on the engagement of multiple sensor monitoring systems for the online acquisition of cutting force, AE, and vibration signals during the machining process. The features extracted after signal processing are employed for the calculation of proper thresholds and the values of relevant parameters (e.g., normalized cutting force, etc.) to be compared with thresholds. As long as the value of a parameter does not exceed the threshold, the signal acquisition proceeds with new updated thresholds; otherwise, a control action should be performed (e.g., the machine is stopped).

In Caggiano et al. (2016), a knowledge-based CTF detection procedure for identifying the occurrence of CTF in turning of Ti-6Al-4V titanium alloy was proposed.

The procedure employed is based on sensor signal acquisition of the three cutting force components (F_x, F_y, and F_z) during machining, following previous results reported in Balsamo et al. (2016). The sensor signals were acquired during experimental turning tests performed under different cutting conditions (varying cutting speed, feed rate, and depth of cut). Sensor signal processing and analysis were performed on the acquired signals and relevant sensor signal statistical features were extracted and employed to develop a methodology for CTF detection. With the aim to automatically identify the occurrence of a CTF during turning, an algorithm was implemented based on the values of the selected statistical features in all the acquired signals.

The methodology was based on the extraction of relevant signal features from signal segments corresponding to 10 ms of machining, consisting of 100 samplings at sampling frequency of 10 kS/s: below this time window, the CTF procedure proved not to be sufficiently reliable. With the aim to speed up the CTF diagnosis, instead of considering consecutive signal segments of 100 samplings each, a moving window with a total length of 100 samplings and a step of 20 samplings was considered. Therefore, at each step, very small signal segments, corresponding to 2 ms of machining, were sequentially processed and collected into larger signal segments corresponding to 10 ms of machining to extract selected signal features (e.g., signal mean, variance, and max–min range) useful for CTF event identification: these features were input to a knowledge-based algorithm comparing the feature values with previously specified thresholds for CTF detection.

The knowledge-based CTF detection procedure operated in the following way: the CTF detection algorithm receives in input the sensor signal segments as well as the specifications concerning the machining process, the tool, and the work material. These data allow identification of the proper procedure for CTF detection and the values of the thresholds for the specific process conditions.

The sensor signal arrays containing 20 samplings of each signal corresponding to 2 ms of cutting time were sequentially buffered and delivered in input to the CTF detection algorithm; the output of the knowledge-based CTF detection algorithm is a Boolean variable, where true means that a CTF event is detected. In the "true" case, an emergency process halting command is fed to the machine tool control.

A statistical approach was employed to detect if there was any cutting force component less effective than the others in detecting a CTF event. The results showed that the component of the cutting force in the z direction, F_z, was generally less robust in CTF detection compared to the F_x and F_y components, as in some cases, the F_z signal component behavior generated a false CTF detection alarm. Therefore, in order to detect the occurrence of CTF, a condition was added to the developed methodology, requiring that at least two out of the three cutting force components signals simultaneously identify a CTF event (Balsamo et al. 2016). Figure 6.5 shows the F_x, F_y, and F_z cutting force components simultaneously detecting a CTF event.

6.3.2.2.3   Sensor monitoring for machinability evaluation   Among the low-machinability titanium alloys, Nitinol, which is a shape-memory nickel-titanium (Ni-Ti) alloy, is particularly difficult to machine as it exhibits severe strain hardening, high toughness and viscosity, as well as a unique pseudoelastic behavior (Weinert et al. 2004). The exceptional strain–stress behavior of these alloys, which tend to strongly harden when deformed, leads to rapid tool wear and is responsible for the development of adverse chip forms and inferior workpiece quality due to adhesions and burrs formation (Guo et al. 2013).

In this framework, a multisensor monitoring procedure based on advanced sensor signals processing and cognitive pattern recognition can provide a valuable support to estimate the machinability of Ni-Ti alloys.

In Segreto et al. (2015), a multisensor monitoring procedure was developed with the aim to allow for a consistent and robust classification of process machinability conditions in turning of Nitinol alloy under

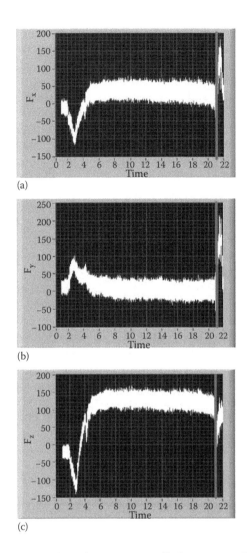

(a)

(b)

(c)

Figure 6.5 Simultaneous CTF detection on all the cutting force components: (a) F_x, (b) F_y, (c) F_z. (From Balsamo, V., Caggiano, A., Jemielniak, K., Kossakowska, J., Nejman, M., Teti, R., *Procedia CIRP*, 41, 939–944, 2016. With permission.)

diverse cutting parameters based on the acquisition of sensor signals relative to cutting force and acceleration of vibrations.

The classification of process conditions was performed on the basis of various quality parameters: tool wear (namely crater and flank wear), vibration level during machining, and type of formed chip.

The multisensor system utilized in the experimental turning tests comprised two diverse sensing units: a Kistler 9265A1 dynamometer for the acquisition of the three cutting force components (F_x, F_y, and F_z, see Figure 6.6a) and a 3D accelerometer bolted on the tool holder in proximity of the tool insert for the acquisition of the three vibration acceleration components (a_x, a_y, and a_z, see Figure 6.6b) with a sampling rate of 2.5 kHz.

The acquired sensorial data were subjected to an advanced sensor signal processing procedure based on signal spectral estimation, allowing for feature extraction from the signal frequency content (Teti et al. 2010; Teti 2015). Sensor fusion technique was applied to combine the information produced by different sensors with the aim to improve the characterization of the machining process conditions. Therefore, the extracted features were employed to construct, on the one hand, feature pattern vectors relative to a single signal component and, on the other hand, SFPVs' combining features coming from the different signals. Both kinds of feature pattern vectors were fed as input to adaptive neuro-fuzzy systems with the aim to assess the Ni-Ti alloy machinability through the identification of correlations between the sensorial data and process acceptability. The neuro-fuzzy system results showed that the greatest pattern recognition performance was achieved by using the SFPV comprising features extracted from all the acquired sensor signals, followed by the a_y and a_z

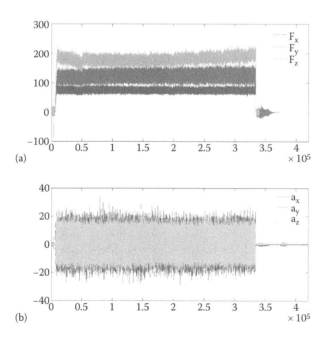

Figure 6.6 Raw signals of: (a) cutting force components (b) vibration acceleration components. (From Segreto, T., Caggiano, A., Karam, S., Teti R., *Sensors*, 17, 2885, 2015. With permission.)

acceleration components and the F_y and F_z force components. The results showed that the y and z signal components were more sensitive than the x signal components, although a general excellent pattern recognition performance was obtained by the neuro-fuzzy models.

6.3.2.2.4   Tool condition monitoring   Kosaraju et al. (2013) presented an online procedure based on the acquisition of AE signals for tool wear prediction in turning of Ti-6Al-4V titanium alloy with physical vapor deposition (PVD)-coated carbide tools. The RMS value of the AE signal at the chip–tool contact was employed to detect the development of the cutting tool flank wear, under different cutting speed, feed and depth of cut values. The flank surface of the cutting tools employed in the machining tests was examined via X-ray spectroscopy technique to identify the wear nature. Using the process parameters like speed, feed, and depth of cut, together with the flank wear values, a mathematical model for predicting the AE signal was developed. With the aim to verify the accuracy of the model, a confirmation test was carried out, showing that the AE signal in Ti alloy turning can be estimated with acceptable accuracy within the considered range of process parameters. The methodology presented by Kosaraju et al. (2013) confirmed that the AE signal can be satisfactorily estimated, allowing monitoring the tool condition during the machining process.

Jie et al. (2008) discussed the basic requirements for TCM based on the acquisition of sensor signals and investigated the effectiveness of two different sensor signals, i.e., cutting force and AE, which are widely employed for machining process monitoring, with the aim to develop a TCM system for turning Ti-6Al-4V titanium alloy. The effectiveness of the selected sensor signals in TCM was investigated based on the features extracted from these signals, considered individually or in combination to form feature sets for effective TCM. From the analysis of the experimental results based on Ti-6Al-4V titanium alloy turning, with acquisition of AE signals (RMS) and cutting force signals, the cutting force resulted as an effective monitoring signal; therefore, a set of features extracted from this signal was employed as input to identify the tool condition in machining of titanium alloy. A suitable sensing technique and a proper set of features for the identification of tool condition were determined, which could also be applied for monitoring other machining processes.

6.3.2.2.5   Surface roughness   Surface roughness is essential to attain the desired product quality and affects fatigue strength, corrosion resistance, precision fits, tribological, and aesthetic requirements. In general, the roughness of a machined surface is evaluated only after the machining process completion, and the parts that fail the surface roughness test need to be reworked with additional cost. Therefore, the possibility to perform an online estimation of surface roughness represents a very interesting issue in the metal machining research area.

In Upadhyay et al. (2013), a methodology based on the employment of vibration signals for the online prediction of surface roughness during turning of Ti-6Al-4V alloy was proposed. To increase the stiffness of machining system, the Ti-6Al-4V alloy workpiece was clamped between a three-jaw chuck and the rotating center of a stiff, high-power precision lathe. Uncoated cemented carbide inserts were employed in the experiments. The study was performed in two stages. At first, only the amplitude of the tool vibration acceleration components in the axial, radial, and tangential directions were employed to develop multiple regression models for predicting the surface roughness, providing a maximum percentage error about 24%. Afterward, correlation analysis was carried out to verify the degree of correlation between cutting speed; feed rate; depth of cut; vibration acceleration amplitude in the axial, radial, and tangential directions; and surface roughness. On the basis of this analysis, feed rate and depth of cut, together with the amplitude of vibration accelerations in the radial and tangential directions, were selected as input parameters to elaborate an advanced first-order multiple regression model for the prediction of surface roughness. The advanced model allowed obtaining a suitable accuracy for surface roughness prediction with a maximum percentage error of 7.45%. Moreover, an ANN model was set up with the aim to integrate it into a computer integrated manufacturing environment.

6.3.2.3   Drilling

6.3.2.3.1   Surface roughness and hole diameter monitoring   In Cruz et al. (2013), a sensor monitoring procedure based on the acquisition of multiple signals during machining and the employment of an ANN was proposed for estimating hole diameter and surface roughness in precision drilling of a sandwich made up of titanium and aluminum alloy plates. Hole diameter and hole wall roughness are affected by variations of the drilling parameters and drill bit wear conditions. The use of a multisensor system mounted on the machine tool enables the drilling process monitoring to detect vibrations, increased electrical power requirements, unbalanced forces, and chip flow resistance. A 3D dynamometer was installed on the tool table of the drilling machine, and a special fixture was assembled on the dynamometer platform to hold the sandwich workpieces, an AE sensor, and an accelerometer. To measure the electric power consumption, a Hall effect sensor was connected to the electric circuit of the machine tool. The experimental drilling tests were performed under nine different cutting conditions by combining three diverse spindle speeds (500, 1000, and 2000 rpm) and three feed rates (22.4, 90, and 250 mm/min). To guarantee the tool integrity at the start of each drilling test, for each cutting condition, a new carbide twist drill was employed to machine 162 consecutive holes. During drilling, the sensor monitoring signals were acquired via a high-frequency DAQ board connected to a computer. The average value of each signal was extracted and used as an input for the ANN, together with the values of the employed spindle speed and feed rate. The trained ANN satisfactorily estimated the hole diameter and hole wall roughness of both titanium and aluminum alloys in all the machining test conditions, demonstrating the applicability of the developed procedure for drilling process monitoring. The ANN successfully generalized the process output without being influenced by instantaneous variations.

6.3.2.3.2 *Run out* In microdrilling processes, one of the most critical issues is represented by run out, which consists in an eccentric motion of the drill generated by the severe centrifugal force due to the high rotational speeds. Run out is responsible for dimensional inaccuracy, damage of the workpiece surface, as well as unexpectedly short tool life and sudden breakage of the cutting tool. The possibility to detect run out through the use of sensor monitoring procedures is therefore much interesting to improve microdrilling processes. In Beruvides et al. (2014), a two-step indirect monitoring system capable of predicting the tool run out from the force signals in a microdrilling process was developed. The proposed method combined the FFT approach for extracting features from the acquired force signals with a model based on ANNs for predicting the process condition using the extracted features as input. The model was trained and tested by using the force signals acquired during microdrilling tests of Ti-6Al-4V titanium alloy executed in a microdrilling machine equipped with a piezoelectric force dynamometer to acquire the three force signal components (F_x, F_y, and F_z). A microdrill with a diameter of 0.1 mm was used, and three different feed rates were applied for each material. The trained ANN model was validated with the remaining data that were not used for the training process, and it was able to detect more than 70% of the run out cases with less than 10% of false detections. The application of such methodology to detect and reduce run out can lead to considerable advantaged in terms of tool life and productivity.

6.3.2.3.3 *Temperature monitoring* To investigate the drilling process of titanium alloys, thermal analysis is crucial. In order to monitor the temperature evolution of carbide tools in Ti-6Al-4V drilling, Lazoglu et al. (2017) proposed an innovative device, called rotary tool temperature (RTT), allowing for cutting tool temperature measurements close to the cutting edge. The device consisted of connectors with cold junction compensations for acquiring signals from six thermocouples, an internal memory for storing the signals during machining, an on-board lithium battery, and an internal clock for the synchronization of the multiple acquired signals. The RTT was integrated to a rotary dynamometer and tool holder for simultaneously measuring temperature, force, and torque signals during the drilling operations. In Lazoglu et al. (2017), such device was employed to acquire the signals coming from two thermocouples. To this aim, the drill bit was adapted by realizing two holes via electrical discharge machining; such holes originated from the back of the flute and terminated close to the drill corner, on the one side, and in the middle of the cutting lip, on the other side of the drill bit. The two wires of the thermocouples connected to the RTT device were inserted inside the lubricant canals and the drill bit holes. Experimental drilling tests on Ti-6Al-4V were carried out under different cutting conditions on a DMC DMG 65 CNC machine tool, and the temperature measurements were employed to validate an analytical thermal model for the estimation of the temperature evolution of carbide tools in Ti-6Al-4V drilling.

6.3.2.4   Other machining processes

6.3.2.4.1   Broaching   Axinte et al. (2004) investigated the opportunity to assess the workpiece surface quality in broaching of Ti-6Al-4V titanium alloy via process monitoring by correlating the quality of the machined surface after broaching and the output signals obtained from multiple sensors, namely, AE, vibration, and cutting forces. Experimental broaching tests were performed on a vertical broaching machine fitted with a multiple sensor monitoring system. The cutting conditions were varied based on an orthogonal array with cutting speed, coolant conditions, and tool settings as factors, and three levels of tool wear were considered. The quality of the machined surface was estimated in terms of geometrical accuracy, burr formation, chatter marks, and surface anomalies. The results showed that the cutting force signals allow perceiving geometrical deviations of the machined profile, burr formation, and to a minor degree, chatter marks. On the other hand, the vibration signals proved to be receptive to chatter marks while the AE signal demonstrated to be effective for the detection of small surface anomalies such as pluckings, laps, and smeared material. Nevertheless, a clear distinction between the different types of surface anomalies was not possible from the analysis of the AE signal. The output signals were analyzed both in the time and frequency domain with the aim to develop suitable techniques for a qualitative as well as quantitative assessment of the machined surface quality.

6.3.2.4.2   Orthogonal cutting   In Nguyen et al. (2017), a low-cost, high-bandwidth, nonintrusive measurement system based on a polyvinylidene fluoride (PVDF) thin film piezoelectric dynamic strain sensor was used to characterize chip segmentation during orthogonal cutting of

Ti-6Al-4V. Specifically, orthogonal cutting experiments using Ti-6Al-4V (grade 5) alloy tubes were performed to evaluate the thin film PVDF sensor's ability to measure the chip segmentation frequency. All tests were performed with a 0° rake angle tool with an uncoated tungsten carbide insert and no cutting fluid was used. Images of the serrated chips were taken to determine the chip segmentation frequency, which in turn was compared to the PVDF sensor's frequency decomposition. The results obtained in the orthogonal cutting tests showed that the PVDF-based sensor system can measure the chip segmentation frequency at different cutting speed and feed values with reasonable accuracy. The accuracy is higher at higher feeds, where the PVDF signal-to-noise ratio is larger. The study also showed that the PVDF sensor can be used in situ to characterize the effects of tool wear on the chip segmentation characteristics. It was found that tool wear causes an increase in the chip segmentation frequency and a reduction in the amplitude of chip serration, captured by the PVDF sensor. The proposed measurement system can be used as a low-cost sensor in orthogonal machining applications to detect the onset of chip segmentation and to identify its characteristics as a function of tool condition (Nguyen et al. 2017).

6.3.2.4.3 Grinding The experimental study presented in Kadivar et al. (2018) investigated the effects of established dressing parameters, i.e., dressing overlap ratio and dressing speed ratio, as well as cutting speed on normal and tangential grinding forces and on the obtained surface roughness in the microgrinding of Ti-6Al-4V with very high dressing overlap ratios. A high-precision five-axis CNC machining center was used for grinding experiments on Ti-6Al-4V block samples using a vitrified-bonded diamond grinding pin with a diameter of 2 mm. To measure the forces and surface roughness, a dynamometer and a tactile surface roughness tester were used, respectively. The results showed that the surface quality, grinding forces, and chip loading of the grinding tool are highly affected by the dressing parameters and can be optimized by changing the overlap ratio and the dressing speed ratio.

6.4 Cloud manufacturing framework for smart sensor monitoring of machining

Cloud manufacturing is recognized as one of the most innovative Key Enabling Technologies for modern manufacturing industry and is claiming increasing attention in manufacturing research (Li et al. 2010; Tao et al. 2011; Xu 2012; Zhang et al. 2012).

The field of machining process monitoring could significantly benefit from the implementation of cloud manufacturing, allowing remotely and dynamically acquiring data on the shop floor via sensors and data acquisition systems as well as performing data analysis in remote, where expert know-how can be made available and shared in the cloud, realizing the service-oriented provision of on-line smart diagnosis and automation functionalities for machining process monitoring.

The implementation of a cloud manufacturing framework represents a remarkable advancement for machining process monitoring, allowing exploiting the cloud capabilities in order to offer real-time diagnosis on process conditions according to a service oriented approach. Introducing sensors and networked communication into the factory strongly supports smart in-process diagnosis as well as the timely activation of adaptive actions based on actual process conditions.

In Caggiano et al. (2016), a cloud manufacturing framework was proposed to realize online process monitoring in the machining of low-machinability materials. The framework is configured with particular reference to TCM in turning of Ti-6Al-4V titanium alloys, which is a critical issue due to the rapid development of tool wear and the unpredictable occurrence of CTF when cutting such low-machinability material. The cloud manufacturing framework for online smart TCM is based on the architecture shown in Figure 6.7.

The cloud resources for computing and service provision are connected to the physical resources such as the machine tool and the sensor system, making up a complex cyber-physical system, which can be defined as a "physical and engineered system whose operations are monitored, coordinated, controlled and integrated by a computing and communication core" (Monostori et al. 2016).

The cloud manufacturing architecture is structured in three layers corresponding to physical resources, local server, and cloud server, respectively. This structure allows sharing the computational effort between different resources, which can be geographically distributed and managed by diverse actors.

The physical resources and the local server are both included in the Factory Network, representing the hardware and software resources available within the production system. On the other hand, the cloud service is Internet based and can be potentially connected inside the boundary of the manufacturing company (private cloud) or outside that boundary (public cloud). At the physical level, a multisensor system including a triaxial force sensor, an AE sensor, and a vibration sensor was mounted on the CNC machine in the proximity of the tool, with the aim to capture useful information for TCM. The multisensor system is utilized for the online acquisition during machining of multiple sensor signals that contain relevant information to use as input for the diagnosis on tool conditions. The computing tasks related to sensor signal preprocessing, which involves signal filtering, amplification, A/D conversion, and segmentation, are assigned to the local server. To support high computing power, the local

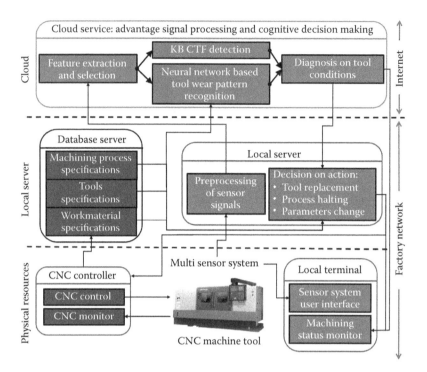

Figure 6.7 Architecture of the cloud-manufacturing framework for smart process monitoring. (From Caggiano, A., Segreto, T., Teti, R., *Procedia CIRP*, 55, 248–253, 2016. With permission.)

machine works as data buffer and preprocesses the data into stand-alone data packages (sensor signal segments), which are delivered via Internet protocols to the on-demand web-based cloud service system according to the needs of the cloud sensor monitoring procedure. The cloud computing capability is employed to rapidly perform online diagnostic tasks, and the potentially huge cloud database is used to maintain and share relevant information and knowledge that can support further cloud services.

The cloud server receives the preprocessed sensor signal segments and performs the diagnostic service consisting of four parts. The first part carries out advanced signal processing on the preprocessed sensor signal segments in order to extract and select relevant sensorial features for TCM. The latter are utilized in two different modules: the first module is employed to recognize CTF events through a knowledge based algorithm, whereas the second module is utilized to estimate consumed tool life through a neural-network-based pattern recognition approach. After the computing task is completed, the cloud sends back the diagnostic output to the local server, which, based on the cloud diagnosis on tool conditions, may select the proper corrective actions. Such actions include process halting, tool replacement, or parameter change: the appropriate command, such as process halting or parameters change, is directly sent from the local server to the CNC control. Moreover, the selected action can be visualized on the local terminal to make the user aware of the control actions taken automatically by the local server. In the case of tool replacement, the local terminal displays a warning to the operator with the instruction to replace the tool.

6.5 Conclusions

To improve and optimize the machining processes applied to the low-machinability titanium alloys, diverse studies in the literature proposed the use of different sensor systems for online monitoring of tool conditions, chip formation, surface integrity, process conditions, chatter detection, etc.

In this chapter, the most relevant sensor monitoring applications regarding the major machining processes applied to titanium alloys were presented. An overview of the main sensors and sensor systems employed for monitoring of machining processes was given. Then, the sensor monitoring applications for each category of machining process, including milling, turning, drilling, broaching, grinding, and orthogonal cutting, were classified based on the specific sensor monitoring scope.

Most of the research works employed cutting force and vibration signals to monitor process conditions, chatter, workpiece quality, and tool wear in diverse machining processes. Also, temperature sensors were employed in a number of cases as the understanding of temperature behavior during machining of titanium alloys represents an important issue to improve cutting process performance because the cutting temperature has a direct effect on both the surface integrity of the workpiece and the machining precision.

In many cases, multiple sensor systems were employed to overcome the limitations of a single sensor and realize the sensor fusion approach integrating information from several sensors of different nature in order to improve the quality and robustness of the process characterization.

The most recent developments in the Industry 4.0 framework, involving the implementation of a cloud manufacturing paradigm to provide smart diagnosis services for online sensor monitoring of machining processes were eventually presented, allowing for enhanced computation and data storage capabilities, available from distributed resources, which can greatly improve process condition diagnosis efficiency and enable more robust decision making by exploiting large information accessibility and knowledge sharing.

References

Armendia, M.; Garay, A.; Villar, A.; Davies, M.A.; Arrazola, P.J. (2010) High bandwidth temperature measurement in interrupted cutting of difficult to machine materials. *CIRP Annals—Manufacturing Technology*, 59(1), pp. 97–100. DOI: 10.1016/j.cirp.2010.03.059.

Arrazola, P.-J.; Garay, A.; Iriarte, L.-M.; Armendia, M.; Marya, S.; Le Maître, F. (2009) Machinability of titanium alloys (Ti6Al4V and Ti555.3). *Journal of Materials Processing Technology*, 209(5), pp. 2223–2230. DOI: 10.1016/j .jmatprotec.2008.06.020.

Axinte, D.A.; Gindy, N.; Fox, K.; Unanue, I. (2004) Process monitoring to assist the workpiece surface quality in machining, *International Journal of Machine Tools & Manufacture*, 44, pp. 1091–1108.

Balsamo, V.; Caggiano, A.; Jemielniak, K.; Kossakowska, J.; Nejman, M.; Teti, R. (2016) Multi sensor signal processing for catastrophic tool failure detection in turning. *Procedia CIRP*, 41, pp. 939–944. DOI: 10.1016/j.procir.2016.01.010.

Beruvides, G.; Quiza, R.; Rivas, M.; Castaño, F.; Habe, R.E. (2014) Online detection of run out in microdrilling of tungsten and titanium alloys, *International Journal of Advanced Manufacturing Technology*, 74(9–12), pp. 1567–1575.

Byrne, G.; Dornfeld, D.; Inasaki, I.; König, W.; Teti, R. (1995) Tool condition monitoring—The status of research and industrial application. *CIRP Annals*, 44(2), pp. 541–567.

Byrne, G.; Dornfeld, D.; Denkena, B. (2004): Advancing cutting technology. *CIRP Annals*, 52(2), pp. 483–507.

Caggiano, A.; Segreto, T.; Teti, R. (2016) Cloud manufacturing framework for smart monitoring of machining. *Procedia CIRP*, 55, pp. 248–253. DOI: 10.1016/j .procir.2016.08.049.

Caggiano, A.; Napolitano, F.; Teti, R. (2017) Dry turning of Ti6Al4V. Tool wear curve reconstruction based on cognitive sensor monitoring. *Procedia CIRP*, 62, pp. 209–214. DOI: 10.1016/j.procir.2017.03.046.

Cruz, C.E.D.; de Aguiar, P.R.; Machado A.R.; Bianchi, E.C.; Contrucci, J.G.; Neto, F.C. (2013) Monitoring in precision metal drilling process using multi-sensors and neural network. *International Journal of Advanced Manufacturing Technology*, 66, pp. 151–158.

Davies, M.A.; Ueda, T.; M'Saoubi, R.; Mullany, B.; Cooke, A.L. (2007) On the Measurement of temperature in material removal processes. *CIRP Annals*, 56(2), pp. 581–604.

Ezugwu, E.O.; Wang, Z.M. (1997) Titanium alloys and their machinability—A review. *Journal of Materials Processing Technology*, 68, pp. 262–274.

Ezugwu, E.O.; Bonney, J.; Yamane, Y. (2003) An overview of the machinability of aeroengine alloys. *Journal of Materials Processing Technology*, 134, pp. 233–253.

Feng, J.; Sun, Z.; Jiang, Z.; Yang, L. (2016) Identification of chatter in milling of Ti-6Al-4V titanium alloy thin-walled workpieces based on cutting force signals and surface topography. *International Journal of Advanced Manufacturing Technology*, 82, pp. 1909–1920.

Guo, Y.; Klink, A.; Fu, C.; Snyder, J. (2013) Machinability and surface integrity of Nitinol shape memory alloy. *CIRP Annals*, 62(1), pp. 83–86.

Huang, P.; Li, J.; Sun, J.; Ge, M. (2012) Milling force vibration analysis in high-speed-milling titanium alloy using variable pitch angle mill. *International Journal of Advanced Manufacturing Technology*, 58, pp. 153–160.

Huang, P.L.; Li, J.F.; Sun, J.; Jia, X.M. (2016) Cutting signals analysis in milling tita-nium alloy thin-part components and non-thin-wall components. *International Journal of Advanced Manufacturing Technology*, 84, pp. 2461–2469.

Ítalo Sette Antonialli, A.; Eduardo Diniz, A.; Pederiva, R. (2010) Vibration analysis of cutting force in titanium alloy milling. *International Journal of Machine Tools and Manufacture*, 50(1), pp. 65–74. DOI: 10.1016/j.ijmachtools.2009.09.006.

Jemielniak, K.; Otman, O. (1998a) Catastrophic tool failure detection based on acoustic emission signal analysis. *CIRP Annals—Manufacturing Technology*, 47(1), pp. 31–34.

Jemielniak, K.; Otman, O. (1998b) Tool failure detection based on analysis of acoustic emission signals. *Journal of Materials Processing Technology*, 76(1–3), pp. 192–197.

Jie, S.; San, W.Y.; Soon, H.G.; Rahman M.; Zhigang W. (2008) Identification of fea-ture set for effective tool condition monitoring – a case study in titanium machining. 4th IEEE Conference on Automation Science and Engineering, Key Bridge Marriott, Washington DC, USA, August 23–26, pp. 273–278.

Kadivar, M.; Azarhoushang, B.; Shamray, S.; Krajnik, P. (2018) The effect of dress-ing parameters on micro-grinding of titanium alloy. *Precision Engineering*, 51, pp. 176–185.

Kosaraju, S.; Anne, V.G.; Popuri, B.B. (2013) Online tool condition monitoring in turning titanium (grade 5) using acoustic emission: Modeling. *International Journal of Advanced Manufacturing Technology*, 67, pp. 1947–1954.

Kurada, S.; Bradley, C. (1997) A review of machine vision sensors for TCM. *Computers in Industry*, 34, pp. 55–72.

Lazoglu, I.; Poulachon, G.; Ramirez, C.; Akmal, M.; Marcon, B.; Rossi, F.; Outeiro, J.C.; Krebs, M. (2017) Thermal analysis in Ti-6Al-4V drilling. *CIRP Annals—Manufacturing Technology*, 66, 105–108

Li, B.H.; Zhang, l.; Wang, S.L.; Tao, F.; Cao, J.W.; Jiang, X.; Song, X.; Chai, X.D. (2010) Cloud manufacturing: A new service-oriented networked manufacturing model. *Computer Integrated Manufacturing Systems* 16(1), pp. 1–7.

Lopez de Lacalle, L.N.; Perez, J.; Llorente, J.I.; Sanchez, J.A. (2000) Advanced cut-ting conditions for the milling of aeronautical alloys. *Journal of Materials Processing Technology*, 100, pp. 1–11.

Monostori, L.; Kádár, B.; Bauernhansl, T.; Kondoh, S.; Kumara, S.; Reinhart, G.; Sauer, O. (2016) Cyber-physical systems in manufacturing. *CIRP Annals—Manufacturing Technology* 65, no. 2, pp. 621–641.

M'Saoubi, R.; Axinte, D.; Soo, S.L.; Nobel, C.; Attia, H.; Kappmeyer, G.; et al. (2015) High performance cutting of advanced aerospace alloys and composite materials. *CIRP Annals—Manufacturing Technology*, 64(2), pp. 557–580. DOI: 10.1016/j.cirp.2015.05.002.

Nguyen, V.; Fernandez-Zelaia, P.; Melkote, S.N. (2017) PVDF sensor based charac-terization of chip segmentation in cutting of Ti-6Al-4V alloy, *CIRP Annals—Manufacturing Technology*, 66(1), pp. 73–76.

Segreto, T.; Caggiano, A.; Teti, R. (2015) Neuro-fuzzy system implementation in multiple sensor monitoring for Ni-Ti alloy machinability evaluation. *Procedia CIRP*, 37, pp. 193–198.

Segreto, T.; Caggiano, A.; Karam, S.; Teti R. (2017) Vibration sensor monitoring of nickel-titanium alloy turning for machinability evaluation. *Sensors*, 17(12), p. 2885.

Sun, Y.; Sun, J.; Li, J.; Xiong, Q. (2014) An experimental investigation of the influence of cutting parameters on cutting temperature in milling Ti6Al4V by applying semi-artificial thermocouple. *International Journal of Advanced Manufacturing Technology*, 70(5–8), pp. 765–773.

Tao, F.; Zhang, L.; Venkatesh, V.C.; Luo, Y.; Cheng, Y. (2011) Cloud manufacturing: A computing and service-oriented manufacturing model. *Proceedings of the Institution of Mechanical Engineers, Part B: Journal of Engineering Manufacture*, 225(10), pp. 1969–1976.

Teti, R. (2015) Advanced IT methods of signal processing and decision making for zero defect manufacturing in machining. *Procedia CIRP*, 28, pp. 3–15.

Teti, R.; Jemielniak, K.; O'Donnell, G.; Dornfeld, D. (2010) Advanced monitoring of machining operations. *CIRP Annals—Manufacturing Technology*, 59(2), pp. 717–739.

Upadhyay, V.; Jain, P.K.; Mehta, N.K. (2013) In-process prediction of surface roughness in turning of Ti-6Al-4V alloy using cutting parameters and vibration signals. *Measurement*, 46, pp. 154–160.

Weinert, K.; Petzoldt, V.; Kötter, D. (2004) Turning and drilling of NiTi shape memory alloys. *CIRP Annals*, 53(1), pp. 65–68.

Xu, X. (2012) From cloud computing to cloud manufacturing. *Robotics and Computer-Integrated Manufacturing*, 28, pp. 75–86.

Zhang, L.; Luo, Y.; Tao, F.; Li, B.H.; Ren, L.; Zhang, X.; Guo, H.; et al. (2012) Cloud manufacturing: A new manufacturing paradigm. *Enterprise Information Systems*, 8(2), pp. 167–187.

chapter seven

Cryogenic machining of titanium alloys

Stefania Bruschi and Andrea Ghiotti

Contents

7.1 General characteristics

Nitrogen is an inert, safe, nontoxic, nonflammable, colorless, and odorless gas, which quickly evaporates and does not leave residues on the machined surfaces, chips, and machine tool components, thus eliminating disposal costs. In general, the following benefits are expected when applying it during machining operations instead of conventional cutting fluids:

- Increase in the tool life, thanks to lower adhesion and diffusion as well as abrasion wear phenomena, the former as a consequence of the low temperatures that inhibit the thermally activated wear phenomena, the latter as a consequence of its lubricity effect.
- Increase in the material removal rate, thanks to the reduction of the tool wear, resulting in higher productivity.
- Improvement in the machined surface integrity in terms of surface quality and mechanical characteristics.
- Enhancement of the machined part functional performances as a consequence of the improved surface integrity.

Uehara and Kumagai (1968) were the first to introduce the term "cryogenic machining," showing an improvement in surface roughness when machining pure titanium. But it was not until the end of the last century that the use of cryogenics fluids was again considered in machining due to the more and more stringent environmental issues related to the disposal of cutting fluids. Furthermore, the introduction of more and more difficult-to-cut alloys to manufacture added-value parts, causing significant tool wear during machining operations, forced to find alternative cutting fluids with an enhanced cooling capacity.

A systematic methodology employing experimental results and analytical techniques was used in Shokrani et al. (2016a) to evaluate the energy consumption in roughing end milling operations when using dry, flood, and cryogenic cooling conditions. Dry and cryogenic cooling techniques were found to be the most favorable environments to minimize power consumption, since, on average, 40% of the power consumption in flood cooling was attributed to the coolant pump. Furthermore, cryogenic cooling allowed increasing seven times the cutting speed and productivity compared to dry cutting.

7.2   Machinability under cryogenic cooling

7.2.1   Tool wear

Up to now, cryogenic cooling has been implemented mainly in turning, especially in case of roughing operations and, to a less extent, in milling. This is due to the intrinsic easiness of adduction of liquid nitrogen to the cutting zone in turning, as the adopted nozzles for the fluid delivery are usually fixed to the tool holder and move with it. The most used configuration includes two nozzles for the liquid nitrogen delivery, one to the rake face and one to the flank face, in order to maximize the fluid performances. On the other hand, in milling, two different approaches in delivery can be used, namely, an external nozzle that is made to move according to the programmed cutting path integral to the cutting tool or a channel inside the mandrel, passing through the cutting tool, which therefore adduces the fluid directly to the cutting zone. While the former is quite easy to be implemented, the latter, even if more accurate, expects a proper design of the mandrel to avoid any thermal damaging as a consequence of the very low temperatures.

As an example, in Park et al. (2014), an external nozzle was used to deliver liquid nitrogen to the cutting zone in case of face milling Ti6Al4V, proving superior machining performances than dry and flood cooling; however, the best results were obtained when implementing cryogenic cooling with Minimum Quantity Lubrication (MQL), as a consequence of the combined lubrication and cooling action.

The effect of internal and external spray methods for liquid nitrogen adduction when carrying out end milling was investigated in Park et al. (2015)

together with hybrid techniques, based on the combination of MQL and cryogenic cooling: a significant reduction of the cutting forces and tool wear was observed when using MQL together with internal cryogenic cooling, especially in the case of deep axial depth-of-cut machining.

Different hybrid cooling/lubricating strategies were also investigated in Sartori et al. (2017a), where it was proved that the adoption of low-temperature coolants on the tool rake face and MQL on the tool flank face helped in significantly reducing the crater and flank wear compared to the sole techniques. A proof of that is given in Figure 7.1, where the two-dimensional (2D) profiles of the tool rake face are presented as a function of the applied technology: the best results were found in the case of the combination of liquid nitrogen and MQL, with the crater wear almost completely eliminated.

In Sun et al. (2010), an alternative cooling approach was adopted, consisting of compressed air cooled by liquid nitrogen delivered to both the rake and flank faces of the tool; a change in the chips characteristics from irregular to regular segmented was proved, with a reduction in the cutting forces compared to dry machining.

In Bermingham et al. (2012), it was proved how the liquid nitrogen delivery method as well as the coolant nozzle position were fundamental to effectively obtain an increased tool life when using cryogenic cooling; on the other hand, the use of high-pressure coolants proved to be less sensitive to the nozzle location compared to the cryogenic coolant.

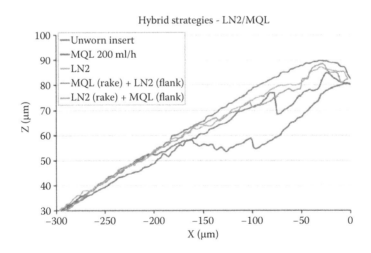

Figure 7.1 2D profiles of the tool rake face when using hybrid lubricating/cooling strategies. (Reprinted from *Wear*, 376, Sartori, S., Ghiotti, A., Bruschi, S., Hybrid lubricating/cooling strategies to reduce the tool wear in finishing turning of difficult-to-cut alloys, 107–114, Copyright 2017, with permission from Elsevier.)

The impact of the pressure and flow rate of the liquid nitrogen on tool life and surface integrity was assessed in Ayed et al. (2017) when machining the Ti6Al4V titanium alloy. The pressure was varied in the range between 4 and 10 bars, and the flow rate between 1.6 and 3.2 l/min, using different nozzle diameters: it was shown that the tool wear and surface integrity strongly depended on the aforementioned parameters, with the best performance obtained at the highest pressure and highest flow rate. This is shown in Figure 7.2, where the evolution of the extension of the flank wear, in case of dry, flood, and cryogenic cooling conditions, is reported. The study demonstrated the need to reconcept the existing tool holders in order to take more advantage of the cryogenic assistance.

Microstructural observations and analyses of the Ti6Al4V chip morphology in Bermingham et al. (2011) suggested that the cryogenic coolant was able to prevent the amount of heat generated and transferred to the tool, as the tool–chip contact length was reduced and, therefore, the generated frictional heat. However, it was stated that only if the optimal machining parameters were used, then the tool life could be improved further thanks to cryogenic cooling.

To assess whether the cryogenic fluid had cooling and/or lubricating capabilities, a specially designed tribometer equipped with a cryogenic injection system was developed in Courbon et al. (2013) to characterize the tribological behavior of Ti6Al4V against uncoated and coated tungsten carbide tools under dry and cryogenic cooling conditions, the latter using both liquid phase nitrogen and gas phase nitrogen. A drastic reduction of the heat transmitted to the pin was found, but almost no effect on the friction coefficient (see Figure 7.3), pointing out that cryogenic cooling could eliminate temperature-induced wear phenomena but did not show a lubricating capability.

On the other hand, in Yousfi et al. (2017), a decrease in the friction coefficient was found when using cryogenic cooling and CrN-coated carbide tools on a self-made tribometer, highlighting the influence of coating thickness and roughness.

Besides the conventional Ti6Al4V titanium alloy, tool wear behavior when cryogenic machining other titanium-based alloys has been recently investigated: the machinability of Ti-5553, a near-beta titanium alloy that is generally considered as a suitable candidate to replace Ti6Al4V in the aerospace industry, was evaluated under cryogenic cooling and compared to that obtained from machining with flood cooling and MQL, showing a reduction of about 30% in the cutting forces, as well as limited nose wear; on the other hand, surface roughness under the MQL condition was better, thanks to the lubricity effects and associated softening of the working material (Sun et al. 2015).

Roughing and finishing operations were carried out on a Ti17 alloy using the cryogenic cooling support in Trabelsi et al. (2017), showing

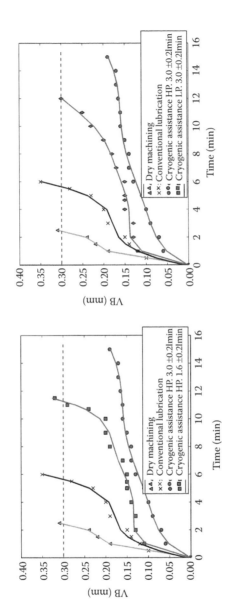

Figure 7.2 Evolution of flank wear as a function of cooling strategy, as well as pressure and flow rate in case of cryogenic cooling. (With kind permission from Springer Science+Business Media: *Int. J. Adv. Manuf. Technol.*, Impact of supply conditions of liquid nitrogen on tool wear and surface integrity when machining the Ti-6Al-4V titanium alloy, 93, 2017, Ayed, Y., Germain, G., Publil Melsio, A., Kowalewski, P., Locufier, D.)

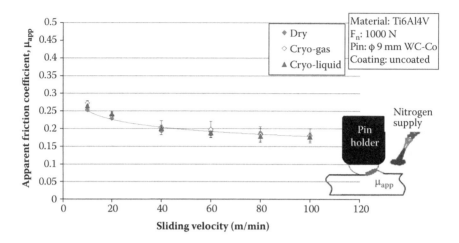

Figure 7.3 Influence of the cryogenic cooling conditions on the apparent friction coefficient when testing Ti6Al4V against uncoated carbide pins. (Reprinted from *Tribol. Int.*, 66, Courbon, C., Pusavec, F., Dumont, F., Rech, J., Kopac, J., Tribological behaviour of Ti6Al4V and Inconel718 under dry and cryogenic conditions—Application to the context of machining with carbide tools, 72–82, Copyright 2013, with permission from Elsevier.)

an improvement in tool life regardless of the cutting parameters, even if at increasing cutting speed, the cryogenic support became lower. Furthermore, if compared with high-pressure water jet support, the benefits of cryogenic cooling were lower in case of finishing operations.

7.2.2   Surface integrity

Different parameters can be used to assess the surface integrity of machined parts, among them, the surface quality, in terms of surface roughness and topography; microstructural alterations/transformations in the machining-affected layer; and mechanical characteristics, in terms of microhardness and residual stresses.

The surface integrity in machining Ti6Al4V under cryogenic cooling conditions was compared to that obtained under dry and MQL in Rotella et al. (2014). The cryogenically machined samples showed a superior surface integrity in terms of surface finish, hardness, and grain size; in particular, it was found that cryogenic cooling improved the dynamic recrystallization mechanism, preventing grain growth after it, without achieving the high beta volume fraction of the dry machined samples. Figure 7.4 reports the effect of cutting speed and feed rate on the mean surface roughness at varying cooling conditions, proving an improvement of the surface finish in case of cryogenic cooling, especially if compared to dry cutting.

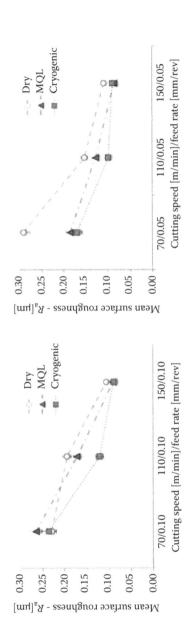

Figure 7.4 Mean surface roughness of samples machined under dry, MQL, and cryogenic cooling at varying cutting speeds and feed rates. (With kind permission from Springer Science+Business Media: *Int. J. Adv. Manuf. Technol.*, The effects of cooling conditions on surface integrity in machining of Ti6Al4V alloy, 71, 2014, 47–55, Rotella, G., Dillon, O.W. Jr., Umbrello, D., Settineri, L., Jawahir, I.S.)

Response surface and artificial neural network methodologies were used in Mia et al. (2017) to correlate the cutting speed and feed rate with the surface roughness when turning under cryogenic cooling using a depth of cut of 1 mm and coated carbide tools: a cutting speed less than 110 m/min was likely to give a favorable machining response, whereas the higher the feed rate, the better the machining performances.

The surface integrity of the Ti6Al7Nb titanium alloy, which is emerging in the biomedical field as a substitute for Ti6Al4V in dental implants and femoral stem prostheses, was evaluated under cryogenic cooling in comparison with dry and flood-cooling machining (Sun et al. 2016): the surface roughness decreased and the hardness increased thanks to the formation of a severe plastic deformation layer with less volume fraction of alpha phase.

A comprehensive investigation of the surface integrity in cryogenic end milling Ti6Al4V was carried out in Shokrani et al. (2016b) using an external nozzle to deliver the liquid nitrogen. Surface roughness was proved to be drastically reduced and a lower amount of surface defects was found (see Figure 7.5). Furthermore, the subsurface microhardness was improved in case of cryogenic cooling, but with a lower extension of the machining-affected layer compared to flood cooling.

On the contrary, in Krishnamurthy et al. (2017), cryogenic cooling proved to increase the surface quality compared to dry cutting but led to

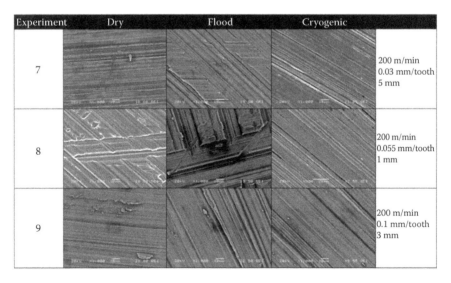

Figure 7.5 Surface defects of samples machined under dry, flood and cryogenic cooling at varying feed rate in end milling. (Reprinted from *J. Manuf. Proc.*, 21, Shokrani, A., Dhokia, V., Newman, T., Investigation of the effects of cryogenic machining on surface integrity in CNC end milling of Ti–6Al–4V titanium alloy, 172–179, Copyright 2016, with permission from Elsevier.)

worse outcomes compared to flood cooling in case of orthogonal cutting tests carried out at a cutting speed lower than in the previous cases (e.g., 45 m/min) (see Figure 7.6); in the same work, the chip formation mechanism and chip segmentation were also evaluated, showing that cryogenic cooling had a more pronounced fracture effect than dry machining, as a consequence of the toughness reduction with the temperature witnessed by Charpy V-notch impact tests. The ease of fracture led to the formation of shorter chip segments and, therefore, reduced cutting forces.

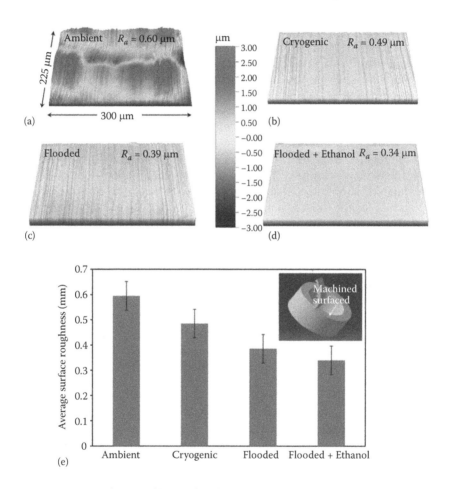

Figure 7.6 3D surface profiles under dry (a), cryogenic (b), and flooded cooling conditions (c) and (d); mean surface roughness under the same conditions (e). (Reprinted from *CIRP J. Manuf. Sci. Technol.*, 18, Krishnamurthy, G., Bhowmick, S., Altenhof, W., Alpas, A.T., Increasing efficiency of Ti-alloy machining by cryogenic cooling and using ethanol in MRF, 159–172, Copyright 2017, with permission from Elsevier.)

7.2.3   Application to additive manufactured titanium alloys

In recent years, Additive Manufacturing (AM) technologies have been introduced for manufacturing biomedical components as they allow producing
customized implants in small batches, with improved biofunctional characteristics. Despite the possibility of obtaining near-net-shape products, some
machining may still be needed on functional surfaces to get the required
tolerances and surface finishes. The different microstructures obtainable on
the basis of the specific AM process may substantially vary the machinability of the investigated AM Ti6Al4V compared to the one of the wrought
alloy. Furthermore, the issue of having clean surfaces becomes even more
challenging in order to reduce the subsequent cleaning steps, which makes
mandatory the avoidance or, at least, the limitation of contaminating cutting fluids during machining. A few works have recently investigated the
impact of cryogenic cooling on finishing machining of AM Ti6Al4V alloys.

A systematic study of the tool wear in cryogenic machining Ti6Al4V
obtained through Electron Beam Melting (EBM) can be found in Bordin
et al. (2015), where it was proved that cryogenic cooling was able to prevent the formation of the wear crater even when the most severe cutting
parameters were adopted. The quality of the machined surfaces was also
investigated, showing sensible improvements with the liquid nitrogen
application for the most severe cutting parameters, thanks to a lower nose
wear (Bordin et al. 2017); cleaner surfaces with less amount of adhered
particles were also found, even if wavier with the presence of jagged feed
marks compared to those under conventional flood cooling. A considerable difference in the chip morphology was also found, as the low ductility induced by the LN2 on the cooled chips favored their breakability, thus
avoiding entanglements around the tool holder (Figure 7.7).

The correlation between the mechanical and thermal characteristics
of the Ti6Al4V obtained through different AM technologies and the tool
wear mechanism was investigated in Sartori et al. (2017b), showing how
the cryogenic cooling prevented the formation of the wear crater on the
tool rake face, with the exception of the direct metal laser sintered alloy,
which presented the highest hardness and lowest thermal conductivity
among the AM variants of the tested alloy and therefore was characterized by the most reduced machinability. On the contrary, the EBM alloy
presented the best machinability, comparable to that of the wrought alloy.

When cryogenic cooling is applied during semi-finishing/finishing
machining operations, the obtainable geometrical accuracy of the machined
components represents one of the major concerns. In Bordin et al. (2016),
a feasibility study was proposed to verify the applicability of cryogenic
cooling when turning EBM Ti6Al4V acetabular cups under dry, wet, and
cryogenic cooling conditions: it was found that cryogenic cooling induced
more thermal contraction on the tool-holder than on the workpiece, and

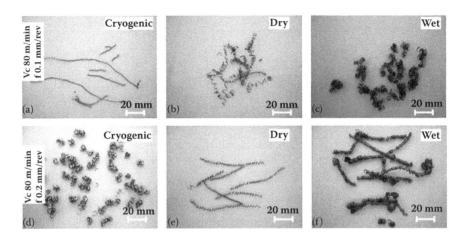

Figure 7.7 Chip morphology under dry (b–c), flood (e–f) and cryogenic (a–d) cooling when turning the EBM Ti6Al4V. (Reprinted from *J. Cleaner Prod.*, 142, Bordin, A., Sartori, S., Bruschi, S., Ghiotti, A, Experimental Investigation on the Feasibility of Dry and Cryogenic Machining as Sustainable Strategies when Turning Ti6Al4V Produced by Additive Manufacturing, 4142–4151, Copyright 2017, with permission from Elsevier.)

that for short cutting times, less than 1 minute, geometrical deviations comparable to those obtained under wet conditions were reached.

7.3   Numerical modelling

Several attempts have been recently made to develop accurate numerical models of cryogenic-assisted machining operations in order to avoid costly and time-consuming experimental trials and provide an effective design support on the basis of the outcomes of numerical sensitivity analyses.

An orthogonal cutting process carried out on Ti6Al4V was modeled in Davoudinejad et al. (2015) using the commercial finite element–based code AdvantaEdge™, providing a sensitivity analysis to the jet radius and heat convection coefficient to reach the appropriate condition to simulate cryogenic cooling. The predicted numerical results were validated through the experimental findings, in terms of cutting forces and chip morphology.

A 3D numerical model of the turning process conducted on Ti6Al4V under dry and cryogenic cooling conditions using the commercial finite element-based code Deform™ was developed in Imbrogno et al. (2017). The numerical model, experimentally calibrated and validated, was able to properly predict the machining-induced microstructural alterations, as the plastic deformed layer that was compared with the same layer estimated by SEM analysis. A user subroutine was implemented in the numerical model of the turning process conducted on the EBM Ti6Al4V, permitting predicting the peculiar workpiece microstructure (alpha

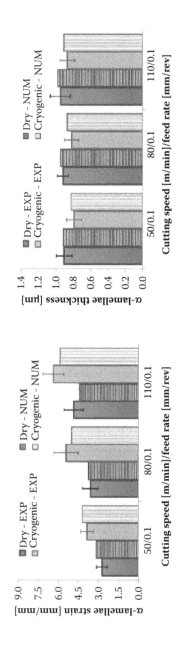

Figure 7.8 Comparison between experimental and predicted strain and thickness of alpha-lamellae in EBM Ti6Al4V turning under cryogenic cooling. (Reprinted from *CIRP J. Manuf. Sci. Technol.*, 18, Umbrello, D., Bordin, A., Imbrogno, S., Bruschi, S., 3D finite element modelling of surface modification in dry and cryogenic machining of EBM Ti6Al4V alloy, 92–100, Copyright 2017, with permission from Elsevier.)

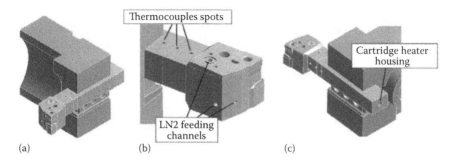

Figure 7.9 Model of a novel concept of tool holder: (a) general overview; (b) thermocouples and LN2 feeding channels location; (c) cartridge heater housing. (Reprinted from *Procedia CIRP*, 58, Novella, M.F., Sartori, S., Bellin, M., Ghiotti, A., Bruschi, S., Modelling the thermomechanical behaviour of a redesigned tool holder to reduce the component geometrical deviations in cryogenic machining, 347–352, Copyright 2017, with permission from Elsevier.)

lamellae strain and thickness, see Figure 7.8) and the nanohardness variations induced by the machining operation under dry and cryogenic cooling conditions (Umbrello et al. 2017).

A numerical model of the up milling process conducted on Ti6Al4V was developed in Ma et al. (2015) using the code AdvantaEdge™ together with a separate heat transfer model developed in Abaqus™ to analyze the heat transfer phenomena before and after the engagement of the tool.

The performance of cryogenic machining may be affected by the cryogenic flow characteristics, namely, the physical phenomena occurring inside the pipe that connects the tank to the nozzle. In Tahri et al. (2017), a computational fluid dynamics model was developed to identify the flow modes and inlet parameters of liquid nitrogen inside internal channels of milling tools, showing that the cooling effectiveness depended on the input flow parameters (pressure, velocity, temperature, etc.) as well as on the pipe geometry.

A new concept of tool holder was proposed in Novella et al. (2017) in order to reduce the effect of the liquid nitrogen very low temperatures on the component's final geometry during semifinishing machining: a thermomechanical model of the new tool holder was developed, calibrated, and validated through industrial trials, showing how embedding cartridge heaters in the tool assembly could provide a heat flow compensating the liquid nitrogen cooling action. Figure 7.9 shows the main features of the developed model.

7.4   Functional performances of the machined components

The use of cryogenic cooling is often reported to increase the functional performances of the machined components, thanks to the enhanced surface integrity that can be achieved with a proper choice of the cutting parameters. An example is represented by biomedical implants made of biocompatible metals (Jawahir et al. 2016b), which, furthermore, benefit the drastic reduction of secondary cleaning steps, thanks to the absence of contaminants left on the machined surface by the cryogenic coolants.

The influence of machining parameters and cooling strategies on the reciprocating sliding wear behavior of wrought and Additive Manufactured Ti6Al4V was investigated under conditions resembling the human body: cryogenic machining proved to be an efficient method to assure a higher degree of adhesive wear compared to conventional machining, and therefore, an improvement of the in vitro implant performances (Bruschi et al. 2016). Similar considerations were introduced in Bertolini et al. (2016), where the fretting corrosion behavior of EBM Ti6Al4V was investigated by replicating the same conditions of the taper interface of a modular hip junction: again, cryogenic machining assured a better performance thanks to enhanced mechanical behavior of the machined surface.

Tribocorrosion tests replicating the human mouth environment were carried out in Bertolini et al. (2017) using wrought and EBM Ti6Al4V samples machined under wet, dry, and cryogenic cooling conditions, and zirconia plates: cryogenic machining enhanced the tribocorrosion characteristics, leading to corrosion potential increase and corrosion current decrease (see Figure 7.10), as well as more adhesive wear than the other techniques. Furthermore, it was proved that the EBM samples presented an in vitro behavior similar to that of the wrought ones.

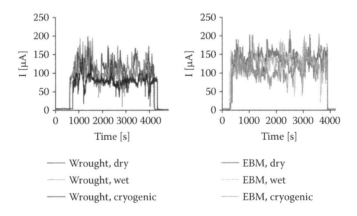

Figure 7.10 Current evolution with time at 0.2 V of applied potential as a function of the machining cooling strategy. (Reprinted from *Biotribology*, 11, Bertolini, R., Bruschi, S., Ghiotti, A., Pezzato, L., Dabalà, M., Influence of the machining cooling strategies on the dental tribocorrosion behaviour of wrought and additive manufactured Ti6Al4V, 60–68, Copyright 2017, with permission from Elsevier.)

7.5   Conclusions

This chapter presented the latest advances in cryogenic machining of titanium alloys, focusing on the tool wear, machined surface integrity, and functional performances of the machined workpiece. Most of the studies refer to the well-known Ti6Al4V alloy, even if recent analyses have been performed on other titanium alloys of aerospace and biomedical interest.

Overall, the use of a cryogenic coolant increases the tool life, both in roughing and finishing operations, and increases the machined surface integrity, especially if compared to dry cutting, but to a less extent if compared to conventional flood cooling. Cryogenic cooling may have also a positive effect on the service life performances of the machined components, as it was demonstrated in the case of biomedical applications in terms of improved wear and corrosion resistance. In general, particular attention must be paid to the design of the machine tool components in order not to damage them as a consequence of the liquid nitrogen low-temperature adduction: to this regard, numerical modelling of the process may be of help in evaluating the thermal fields arising during the cutting operations.

References

Ayed Y, Germain G, Publil Melsio A, Kowalewski P, Locufier D (2017) Impact of supply conditions of liquid nitrogen on tool wear and surface integrity when machining the Ti-6Al-4V titanium alloy. *International Journal of Advanced Manufacturing Technology* 93/1–4, 1199–1206.

Bermingham MJ, Kirsch J, Sun S, Palanisamy S, Dargusch MS (2011) New observations on tool life, cutting forces and chip morphology in cryogenic machining Ti-6Al-4V. *International Journal of Machine Tools & Manufacture* 51, 500–511.

Bermingham MJ, Palanisamy S, Kent D, Dargusch MS (2012) A comparison of cryogenic and high pressure emulsion cooling technologies on tool life and chip morphology in Ti-6Al-4V cutting. *Journal of Materials Processing Technology* 212, 752–765.

Bertolini R, Bruschi S, Bordin A, Ghiotti A, Pezzato L, Dabalà M (2016) Fretting corrosion behaviour of additive manufactured and cryogenically-machined Ti6Al4V for biomedical applications. *Advanced Engineering Materials* 19/6, 1–9.

Bertolini R, Bruschi S, Ghiotti A, Pezzato L, Dabalà M (2017) Influence of the machining cooling strategies on the dental tribocorrosion behaviour of wrought and additive manufactured Ti6Al4V. *Biotribology* 11, 60–68.

Bordin A, Bruschi S, Ghiotti A, Bariani PF (2015) Analysis of tool wear in cryogenic machining of additive manufactured Ti6Al4V alloy. *Wear* 328–329, 89–99.

Bordin A, Medeossi F, Ghiotti A, Bruschi S, Savio E, Facchini L, Bucciotti F (2016) Feasibility of cryogenic cooling in finishing turning of acetabular cups made of Additive Manufactured Ti6Al4V. *Procedia CIRP* 46, 615–618.

Bordin A, Sartori S, Bruschi S, Ghiotti A (2017) Experimental investigation on the feasibility of dry and cryogenic machining as sustainable strategies when turning Ti6Al4V produced by Additive Manufacturing. *Journal of Cleaner Production* 142, 4142–4151.

Bruschi s, Bertolini R, Bordin A, Medea F, Ghiotti A (2016) Influence of the machining parameters and cooling strategies on the wear behaviour of wrought and additive manufactured Ti6Al4V for biomedical applications. *Tribology International* 102, 133–142.

Courbon C, Pusavec F, Dumont F, Rech J, Kopac J (2013) Tribological behaviour of Ti6Al4V and Inconel718 under dry and cryogenic conditions—Application to the context of machining with carbide tools. *Tribology International* 66, 72–82.

Davoudinejad A, Chiappini E, Tirelli S, Annoni M, Strano M (2015) Finite element simulation and validation of chip formation and cutting forces in dry and cryogenic cutting of Ti-6Al-4V. *Procedia Manufacturing* 1, 728–739.

Imbrogno S, Sartori S, Bordin A, Bruschi S, Umbrello D (2017) Machining simulation of Ti6Al4V under dry and cryogenic conditions. *Procedia CIRP* 58, 475–480.

Jawahir IS, Attia A, Biermann D, Duflou J, Klocke F, Meyer D, Newman ST, Pusavec F, Putz M, Rech J, Schulze V, Umbrello D (2016a) Cryogenic manufacturing processes. *CIRP Annals—Manufacturing Technology* 65/2, 713–736.

Jawahir IS, Puleo DA, Schoop J (2016b) Cryogenic machining of biomedical implant materials for improved functional performance, life and sustainability. *Procedia CIRP* 46, 7–14.

Krishnamurthy G, Bhowmick S, Altenhof W, Alpas AT (2017) Increasing efficiency of Ti-alloy machining by cryogenic cooling and using ethanol in MRF. *CIRP Journal of Manufacturing Science and Technology* 18, 159–172.

Ma J, Andrus P, Condoor S, Lei S (2015) Numerical investigation of effects of cutting conditions and cooling schemes on tool performance in up milling of Ti-6Al-4V alloy. *International Journal of Advanced Manufacturing Technology* 78, 361–383.

Mia M, Khan MA, Dhar NR (2017) Study of surface roughness and cutting forces using ANN, RSM, and ANOVA in turning of Ti-6Al-4V under cryogenic jets applied at flank and rake faces of coated WC tool. *International Journal of Advanced Manufacturing Technology* 93/1-4, 975–991.

Novella MF, Sartori S, Bellin M, Ghiotti A, Bruschi S (2017) Modelling the thermomechanical behaviour of a redesigned tool holder to reduce the component geometrical deviations in cryogenic machining. *Procedia CIRP* 58, 347–352.

Park KH, Yang GD, Lee MG, Jeong H, Lee SW, Lee DY (2014) Eco-friendly face milling of titanium alloy. *International Journal of Precision Engineering and Manufacturing* 15/6, 1159–1164.

Park KH, Yang GD, Suhaimi MA, Lee DY, Kim TG, Kim DW, Lee SW (2015) The effect of cryogenic cooling and minimum quantity lubrication on end milling of titanium alloy Ti-6Al-4V. *Journal of Mechanical Sciences and Technology* 29/12, 5121–5126.

Rotella G, Dillon Jr OW, Umbrello D, Settineri L, Jawahir IS (2014) The effects of cooling conditions on surface integrity in machining Ti6Al4V alloy. *International Journal of Advanced Manufacturing Technology* 71, 47–55.

Sartori S, Ghiotti A, Bruschi S (2017a) Hybrid lubricating/cooling strategies to reduce the tool wear in finishing turning of difficult-to-cut alloys. *Wear* 376–277, 107–114.

Sartori S, Moro L, Ghiotti A, Bruschi S (2017b) On the tool wear mechanisms in dry and cryogenic turning of Additive Manufacture titanium alloys. *Tribology International* 105, 264–273.

Shokrani A, Dhokia V, Newman ST (2016a) Energy conscious cryogenic machining of Ti-6al-4V titanium alloy. *Proc. IMech Part B: Journal of Engineering Manufacture* 1–17.

Shokrani A, Dhokia V, Newman T (2016b) Investigation of the effects of cryogenic machining on surface integrity in CNC end milling of Ti-6Al-4V titanium alloy. *Journal of Manufacturing Processes* 21, 172–179.

Sun S, Brandt M, Dargusch MS (2010) Machining Ti-6Al-4V alloy with cryogenic compressed air cooling. *International Journal of Machine Tools & Manufacture* 50, 933–942.

Sun Y, Huang B, Puleo DA, Jawahir IS (2015) Enhanced machinability of Ti-5553 alloy from cryogenic machining comparison with MQL and flood-cooled machining and modeling. *Procedia CIRP* 31, 477–482.

Sun Y, Huang B, Puleo DA, Schoop J, Jawahir IS (2016) Improved surface integrity from cryogenic machining of Ti-6Al-7Nb alloy for biomedical applications. *Procedia CIRP* 45, 63–66.

Tahri C, Laquien P, Outeiro JC, Poulachon G (2017) CFD simulation and optimize of LN2 flow inside channels used for cryogenic machining: Application to milling of titanium alloy Ti-6Al-4V. *Procedia CIRP* 58, 584–589.

Trabelsi S, Morel A, Germain G, Bouaziz Z (2017) Tool wear and cutting forces under cryogenic machining of titanium alloy (Ti17). *International Journal of Advanced Manufacturing Technology* 91, 1493–1505.

Uehara K, Kumagai S (1968) Chip formation, surface roughness and cutting force in cryogenic machining. *CIRP Annals—Manufacturing Technology* 17/1, 409–416.

Umbrello D, Bordin A, Imbrogno S, Bruschi S (2017) 3D finite element modelling of surface modification in dry and cryogenic machining of EBM Ti6Al4V alloy. *CIRP Journal of Manufacturing Science and Technology* 18, 92–100.

Yousfi M, Outeiro JC, Nouveau C, Marcon B, Zouhair B (2017) Tribological behaviour of PVD hard coated cutting tools under cryogenic cooling conditions. *Procedia CIRP* 58, 561–565.

chapter eight

Advanced technologies in drilling of light alloys and CFRP hybrid stacks for airframe structure manufacturing in the aerospace industry

Luigi Nele and Roberto Teti

Contents

8.1   Introduction

During the 1990s, newly established low-cost airlines, i.e., airline companies offering reduced fare flights by elimination of several passenger services, were introduced worldwide and have represented the engine and the fuel of the extraordinary increase in number of flight passengers constantly recorded starting from that period. In fact, from 1990 to 1999, the number of passengers carried by the airline companies has increased from 1.025 billion to 1.562 billion, and up to 3.696 billion in 2016, as ICAO estimates.

Economic and market analyses as reported in Airbus Global Market Forecast 2012–2031 (Airbus 2012) have shown that by 2031, the fleet of passenger and cargo aircrafts, with more than 100 seats and weighing more than 10 tons, will amount to 35,490 aircrafts, thus more than doubling the number of 17,170 aircrafts in service in 2011. Moreover, according to Airbus, by 2031, the world's airlines will deliver more than 28,200 new passenger and cargo aircrafts, for a total value of US$3.96 trillion at 2011 list prices. Concurrently, the number of aircraft manufacturers has increased for the technological advancement of Asian countries and other emerging industrial areas in the world.

This scenario requires aircraft manufacturers to increase their level of competitiveness by designing and producing more efficient aircrafts, e.g., through higher utilization of advanced materials for aircraft structural components and appropriate development of competitive manufacturing technologies (Buhl 1992; Younossi et al. 2001).

All major aircraft manufacturing companies consider that the introduction of composite materials in the entire aircraft fuselage, once the related manufacturing technologies will have been fully developed, could allow for simplified production procedures and better aircraft performance in terms of weight, fatigue resistance, and manufacturing costs, where the main criticality remains the high manufacturing costs (Jerome 2001).

Indeed, although it is well established that the utilization of composite materials provides weight savings that lead to lower operating costs for airlines, the real advantage in developing appropriate composite materials manufacturing technologies for the aviation industry is represented by the opportunity for integrated manufacturing of aircraft parts and consequent lowering of manufacturing costs, which can bring forth significantly higher economic efficiency (Smith 2013).

The technological development of aeronautical structures follows two main lines: the first line covers the development and use of innovative materials and the second line concerns the development of more integrated and more efficient manufacturing technologies.

Carbon-fiber reinforced plastics (CFRPs) appear to be the most suitable high-performance composite materials for the implementation of integrated manufacturing technologies, and several aircraft manufacturing companies have chosen to seriously engage them in the realization of composite-based airframe structures. For example, at the beginning of the third millennium, Boeing started the development of a 787 CFRP fuselage aircraft, currently in service with several carriers worldwide. The materials used in the advanced airframe structures are shown in Figure 8.1 (Owen et al. 2014).

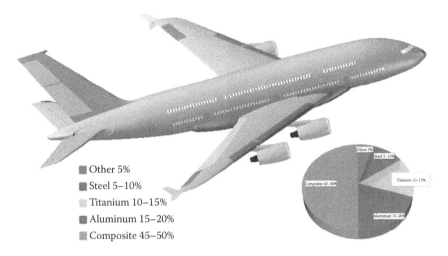

Other 5%
Steel 5–10%
Titanium 10–15%
Aluminum 15–20%
Composite 45–50%

Figure 8.1 Materials used in advanced aircraft structures. (Data from Owen, D., Gardner, S., Modrzejewski, B., Fetty, J., Karg, K., Improving wear and fretting characteristics with fiber reinforced aluminum liners, in: Proceedings of AHS 70th Annual Forum, Montréal, Québec, Canada, vol. 4, pp. 2597–2606, 2014.)

It must be noted that, even in the case of a very innovative aircraft structure such as the Boeing 787, a significant number of light and special alloys must be employed as well, which implicates that appropriate assembly technologies for joining the various parts must be developed.

8.2 Materials for aeronautical structures

In the realization of aeronautical structures, stiffness and strength remain the most significant properties for their materials, although from the 1950s onward, fatigue strength, corrosion resistance, and damage tolerance became indispensable features in the selection of structural materials. More recently, i.e., since the 1970s, new requirements emerged as a result of the evolution of the concepts of safety for structures and of tolerance to impacts of diverse nature to which the aircraft may be subjected during its operational life (Ye et al. 2005; Heimbs et al. 2008). The introduction in the aeronautical field of light alloys such as aluminum alloys and titanium alloys and the development of high-performance composite materials have ensured the achievement of property-to-weight ratio and tolerance to damage so high as to make these materials the most suitable for utilization in the aeronautical industry (Hoskin and Baker 1986). At present, the materials used in the manufacture of airframe structures are the following:

- Ti alloys, mainly Ti4Al6V
- Al alloys, mainly 2024 and 7075
- Composites materials, mainly CFRP

Titanium alloys possess a very attractive combination of properties like high strength-to-weight ratio, low density, and excellent corrosion resistance. Titanium alloys can exist in α, β, or $\alpha - \beta$ phases. In airframe structure manufacturing, the most employed alloy is Ti6Al4V with thickness up to 6 mm and the most common machining operation is drilling (Immarigeon et al. 1995).

Unalloyed titanium (commercially pure) displays a hexagonal close-packed crystal structure, referred to as alpha (α) phase, up to 883°C. At this temperature, the titanium crystal structure transforms to a body-centered cubic structure, known as beta (β) phase. The alloying additions and thermomechanical processing lead to a wide range of alloy types and properties. Based on the phases present in the alloy, titanium alloys can be classified as either α alloys, β alloys, or $\alpha - \beta$ alloys.

Among the titanium alloys, the $\alpha - \beta$ Ti6Al4V is the most widely employed in airframe manufacturing.

Aluminum addition stabilizes and strengthens the α phase, increases $\alpha + \beta \leftrightarrow \beta$ transformation temperature, and reduces alloy density. Vanadium—β-stabilizer—reduces $\alpha + \beta \leftrightarrow \beta$ transformation temperature and facilitates hot working. Depending on the required mechanical properties, heat treatments can be applied to the Ti-6Al-4V alloy.

In Table 8.1, the composition and characteristics of the Ti6Al4V titanium alloy are reported (Welsch et al. 1993). The low thermal conductivity of the titanium Ti6Al4V alloy is a disadvantage in mechanical machining because it causes high heating of the cutting tool.

The positive redox potential implies that if coupled with another material, the titanium alloy tends to reduce instead of oxidize; this feature makes it particularly suitable for mating with composite materials.

Aluminum alloys have been the main airframe materials since they replaced wood starting from the late 1920s. Even though the utilization of composite materials in airframe structures is increasing, the role of aluminum alloys in aircraft fabrication remains important.

Table 8.1 Titanium Ti6Al4V alloy composition and characteristics

		Alloying elements content, wt%				
Al	Mo	V	Cr	Fe	C	Si	Ti
6.1	–	4.3	–	0.16	0.01	–	bal

Mechanical properties		
Ultimate tensile strength	Rockwell hardness	Poisson ratio	Elongation at break	Fatigue strength at 600 MPa	Modulus of elasticity
970 (MPa)	32 (HCR)	0.34	15%	>10,000,000 cycles	120 (GPa)
Yield strength					
930 (MPa)					

Thermal and electrical properties		
Melting temperature	Thermal expansion coefficient	Specific heat	Thermal conductivity	Electrical resistivity	Electrical conductivity
1650 (°C)	8.6×10^{-6} (1/°C)	565 (J/kg°C)	22.7 (W/m²°C)	1.75×10^{-6} (Ωm)	0.9 (%IACS)
Density					
4400 (kg/m³)					

Source: Welsch, G., Boyer, R., Collings, E.W., eds., *Materials Properties Handbook: Titanium Alloys*, ASM International, 1993.

For wrought aluminum alloys, a four-digit system is used to produce a list of wrought composition families as follows (Rooy 1990):

1. Controlled unalloyed (pure) compositions
2. Alloys in which copper is the principal alloying element, although other elements, notably, magnesium, may be specified
3. Alloys in which manganese is the principal alloying element
4. Alloys in which silicon is the principal alloying element
5. Alloys in which magnesium is the principal alloying element
6. Alloys in which magnesium and silicon are principal alloying elements
7. Alloys in which zinc is the principal alloying element, but other elements such as copper, magnesium, chromium, and zirconium may be specified
8. Alloys including tin and some lithium compositions characterizing miscellaneous compositions

The aluminum alloys most used in the manufacture of airframe structures are 2024 and 7075. In particular, the 2024 alloy is used for skins, and the 7075 alloy, for airframe elements.

In Table 8.2, the 2024 and 7075 alloy composition and characteristics are reported (Bray 1990; Cayless 1990).

A composite material is a combination of two or more distinct materials, also named phases. Given the vast range of materials that may be considered as composites according to this definition, it is necessary to specify which are the composite materials presently utilized for the realization of aeronautical structures. The composite materials presently used in the aeronautical field are a combination of two phases, one acting as reinforcement and the other as matrix. The reinforcement is in the form of fibers, which are two-dimensional elements of diameter up to 50 μm and length thousands of times the diameter, and the matrix is the continuous phase that incorporated the reinforcing fibers and transfers the load to them.

Carbon fibers and polymer matrices, in particular epoxy resins, are used in the manufacture of aeronautical structures; this kind of composite material is called carbon-fiber reinforced plastic (CFRP).

In Table 8.3, the characteristics of CFRP and its components are reported (Kelly 1994).

CFRP composites show excellent specific properties (properties per unit weight) allowing for considerable weight savings in aeronautical structures. Moreover, the possibility of manufacturing integrated structures with fewer components is of extreme interest for manufacturing cost reduction.

Table 8.2 2024 and 7075 Aluminum alloys composition and characteristics

Alloy		Alloying element content, wt%						
	Mn	Mg	Cr	Fe	Cu	Si	Ti	Zn
	0.9	1.8	0.10	0.50	4.9	0.50	0.15	0.25

	Mechanical properties				
	Yield strength	Ultimate tensile strength	Rockwell hardness	Poisson ratio	Elogation at break	Fatigue strength	Modulus of elasticity
2024	345 (MPa)	483 (MPa)	46.8 HCR	0.33	15%	138 MPa at 5×10^8 cycles	73.1 GPa

	Thermal and electrical properties				
	Density	Melting temperature	Thermal expansion coefficient	Specific heat	Thermal conductivity	Electrical resistivity	Electrical conductivity
	2780 (kg/m³)	500–640 (°C)	22.9×10^{-6} (1/°C)	875 (J/kg°C)	110–120 (W/m°C)	5.82×10^{-8} (Ωm)	17 (m/Ωmm²)

(Continued)

Table 8.2 (Continued) 2024 and 7075 Aluminum alloys composition and characteristics

Alloy	Alloying elements content, wt%							
	Mn	Mg	Cr	Fe	Cu	Si	Ti	Zn
	0.3	2.5	0.2	0.50	2	0.40	0.15	5.1

Mechanical properties

	Yield strength	Ultimate tensile strength	Rockwell hardness	Poisson ratio	Elogation at break	Fatigue strength	Modulus of elasticity
7075	503 (MPa)	572 (MPa)	53.5 HCR	0.33	10%	159 MPa at 5×10^8 cycles	71.7 GPa

Thermal and electrical properties

Density	Melting temperature	Thermal expansion coefficient	Specific heat	Thermal conductivity	Electrical resistivity	Electrical conductivity
2810 (kg/m³)	574–635 (°C)	23.4×10^{-6} (1/°C)	960 (J/kg°C)	130 (W/m°C)	5.15×10^{-8} (Ωm)	19 (m/Ωmm²)

Source: Cayless, R.B.C., "Alloy and Temper Designation Systems for Aluminum and Aluminum Alloys," in ASM Metals handbook, Vol. II, Properties and Selection: Non Ferrous Alloys and Special-Purpose Materials, 10th edition, 1990; Bray, J.W., "Aluminum Mill and Engineered Wrought Products," in ASM Metals Handbook, Vol. II, Properties and Selection: Non Ferrous Alloys and Special-Purpose Materials, 10th Edition, 1990.

Table 8.3 CFRP characteristics

Polymer	Density (g/mm³)	Ultimate tensile strength (MPa)	Modulus of elasticity (GPa)	Elongation at break (%)	T_{MAX} service (°C)	Thermal conductivity (W/m°C)
Epoxy	1.15	50	4	>2	<100	0.5
Carbon fiber	1.80	4000	220	<2	–	14
Composite UD Vf 50	1.48	1500	105	1.1	<100	–

Source: Kelly, A., ed., *Concise Encyclopedia of Composite Materials*, Pergamon, 1994.

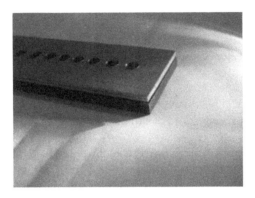

Figure 8.2 Ti-CFRP hybrid stack.

At present, although the major airframe producers have developed technologies to manufacture CFRP fuselage barrels incorporating longitudinal reinforcing stringers, aeronautical structures are still realized by assembling components made of both metal alloys and composite materials, called hybrid stacks of CFRP, Al alloys, and Ti alloys (Figure 8.2) (Terracciano 2012).

8.3   Assembly processes in the aeronautical industry

The manufacturing of airframe structures involves the fabrication of many parts and subassemblies by means of both traditional and innovative manufacturing technologies and the subsequent parts assembly.

The number of fasteners in a typical military aircraft varies from 200,000 to 300,000, while in a civil aircraft, from 1,500,000 to 3,000,000 (Terracciano 2012); for all these fasteners, installation holes are required.

Although the technological evolution of joining processes, such as bonding and welding, has resulted in a drastic improvement in the quality of joints made with these techniques, to date, the assembly of the components that make up the aircraft fuselage is still based on drilling and mechanical fastening of the components to join.

The possibility of realizing fuselage barrels, made of composite materials by integrating longitudinal stringers with external panels, could reduce the drilling and fastening operations. Figure 8.3 shows a fuselage barrel realized in CFRP that integrates longitudinal stringers and external panels.

However, drilling remains a very important process in the manufacture of aeronautical structures, particularly for stringer and panel fastening, from both the quantitative and the qualitative viewpoints, and accordingly, it is widely studied and constantly improved.

The use of lightweight metal alloys, such as aluminum alloys and titanium alloys, in the fabrication of stiffening elements has made the drilling process more critical because, on the assembly lines, hybrid stacks of composite material laminates and metal alloy sheets must be drilled together at the same time.

In the aeronautics industry, the technological evolution of drilling processes, with the development of improved geometry twist drills and drill bits made of high-performance materials, was not accompanied by an equivalent evolution of assembly systems. The assembly of aeronautical structures is still a procedure characterized by high manpower utilization and low automation rate, although automated processing and control

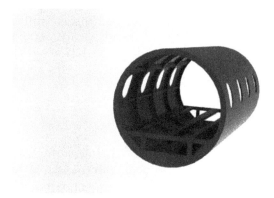

Figure 8.3 CFRP fuselage with integrated longitudinal stringers.

systems have been introduced at major aircraft manufacturers since the end of the second millennium.

As reported, Boeing initially launched the moving assembly line on its B737 final assembly plant (Cort 2008), and Spirit Aerosystems introduced highly automated assembling processes applied to major subassemblies for the B737 and B787 and various Airbus products (Weber 2009).

However, final assembly, subassembly, and single part manufacturing in the aircraft industry remain, to date, high-touch operations.

At present, in the aviation industry, the manufacturing technologies adopted for parts realization do not allow for such a dimensional precision as to carry out the drilling operations off the assembly jigs, whereby it results in necessary drilling of the parts on the assembly lines.

A typical assembly flow for the fuselage structure involves the following steps:

- Positioning of parts or subassemblies on the mounting jigs
- Drilling of the components
- Installation of treated fasteners and sealing
- Cleaning

Due to the dimensions of the parts to be realized and the tight tolerances imposed, proper mounting jigs are used for the correct positioning of the components. The subassemblies are located and positioned on the mounting jigs starting from the implementation of the "coordination holes," the accuracy of which depends on the tolerances of the entire structure. Further parts can then be positioned on such subassemblies to obtain the final structure. In Figure 8.4, an assembly jig for rear plane structure is showed.

Figure 8.4 Assembly jigs with coordination holes.

8.4   Drilling technologies for lightweight sheet materials in aircraft industry

Conventional drilling is a machining process that allows, through the removal of the workmaterial in the form of chips, to obtain holes. This process is generally followed by other operations such as reaming and countersinking.

Although several technologies have been developed for making holes, conventional drilling by means of mechanical tools, called twist drills or drill bits, remains the most widely used hole making process. Choosing the most appropriate tools and methods for making holes depends on the type of workmaterial, hole size, number of holes, and time to complete the operation.

In conventional drilling, the rotary cutting motion is provided by rotating the drilling tool around its axis, whereas the rectilinear feed motion in the axial direction can be given to the tool or the workpiece.

The drill bit typically consists of a cylindrical element into which two, or more, helical grooves are cut, forming cutting lips at the end of the drill bit; a central horizontal chisel edge is present at the drill bit axis to connect the cutting lips.

The twist drill is the most widely used tool. The twist drill is made of two helical grooves forming two cutting lips at the end surface, AB and EC. In Figure 8.5, the scheme of a conventional twist drill is shown.

The helical grooves allow for the evacuation of the chips formed at the cutting edges during drilling. Figure 8.6 reports the twist drill nomenclature, while Figure 8.7 shows the details of the characteristic angles of the twist drill: rake angle, relief angle, and edge angle.

The material removal process during drilling occurs along the cutting lips and at the chisel edge. A force acting in the feed direction, called thrust force, is generated, which is the sum of several components related to the cutting process and the friction between tool and work material. A torque is also generated as a consequence of the cutting process and the friction against the work material.

The cutting component of the torque is due to the cutting force component normal to the feed direction. The frictional component of the torque is caused by the friction between the chisel edge and the work material, the friction between the side surface of the drill and the inside hole surface, and the friction of the flowing chip.

The thrust force and the torque depend on the workpiece materials and geometry, the tool material and geometry, and the ratio between feed rate and cutting speed.

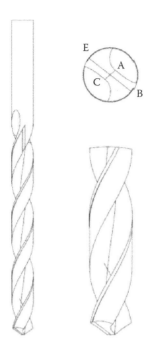

Figure 8.5 Twist drill geometry.

These parameters play an important role in determining both the quality characteristics of the drilled holes and the development of tool wear.

One of the main limitations when drilling light alloy/CFRP stacks with conventional twist drills is represented by the rapid wear experienced by the tool, sometimes after just a few holes.

This rapid tool wear is due both to the abrasive nature of the CFRP reinforcing fibers and to the characteristics of the light alloys.

Tool wear has a critical influence on hole quality and surface integrity as the thrust force increases notably with growing tool wear.

In industrial practice, drilling of light alloy/CFRP stacks is performed as a one-shot operation, either starting from the metal alloy side or from the CFRP composite side of the stack. Beginning from the metal alloy side prevents damage on the hole walls generated in the CFRP by the metal chips and provides for more effective heat dissipation. On the other hand, the onset of exit delamination in the CFRP laminate is more frequent in this case.

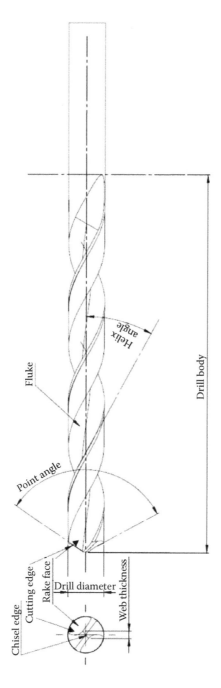

Chisel edge

Cutting edge

Rake face

Drill diameter

Web thickness

Point angle

Fluke

Helix angle

Drill body

Figure 8.6 Twist drill nomenclature.

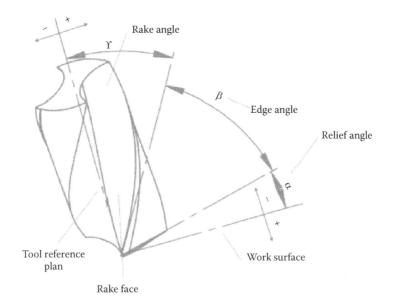

Figure 8.7 Twist drill cutting angles.

8.5   *Quality requirements*

The drilling process in the aeronautical sector is subject to particularly strict limitations, from both the geometric and the dimensional point of views.

The characteristics to be respected are as follows:

- Diameter: in airframe assembly, the dimensional tolerance varies as a function of hole diameter (Table 8.4).
- Angularity: the maximum angular deviation of the hole axis cannot exceed +–2°.

Table 8.4 Dimensional tolerances for drilling in airframe assembling

Nominal diameter, inch (mm)	Structural applications, inch (mm)	
	Min.	Max.
0.1890 (4.801)	0.1895 (4.813)	0.1915 (4.864)
0.2490 (6.325)	0.2495 (6.337)	0.2515 (6.388)
0.3115 (7.912)	0.3120 (7.925)	0.3140 (8.636)
0.3740 (9.500)	0.3745 (9.512)	0.3765 (9.563)

Source: Terracciano, R., *Caratterizzazione della Lavorazione di Foratura di Stacks Ibridi Composito Metallo*, MS thesis, Department of Materials and Production Engineering, University of Naples "Federico II," 2012.

Table 8.5 Delamination extension limits in drilling CFRP

Hole size (inch)	Carbon fabric		Carbon UD	
	"D" Max, inch (mm)	"W" Max, inch (mm)	"D" Max, inch (mm)	"W" Max, inch (mm)
3/16	0.007 (0.178)	0.030 (0.762)	0.014 (0.356)	0.100 (2.54)
1/4	0.007 (0.178)	0.040 (1.016)	0.014 (0.356)	0.100 (2.54)
5/16	0.007 (0.178)	0.040 (1.016)	0.014 (0.356)	0.120 (3.05)
3/8	0.007 (0.178)	0.040 (1.016)	0.014 (0.356)	0.120 (3.05)

- Burrs: the height of the burr in metal alloy drilling cannot exceed the limits that depend on the work material and thickness.
- Delamination in CFRP: the delamination size cannot exceed the limits that depend on the work material, the hole diameter, and the laminate thickness (Table 8.5).

The dimensional tolerances are reported in Table 8.4 (Terracciano 2012)

The delamination extension limits for CFRP are reported in Table 8.5 (Terracciano 2012)

Delamination is the main damage mode in composite drilling and consists of the local in-plane separation between laminae in the laminate thickness around the hole (Ho-Cheng and Dharan 1990).

8.6   Drilling of Al/CFRP and Ti/CFRP hybrid sheet materials in the aerospace industry

In this section, the experimental results obtained during an extensive campaign of drilling tests carried out on Al/CFRP and Ti/CFRP stacks for airframe manufacturing are reported and discussed (Terracciano 2012).

The CFRP laminate is composed of 26 CFRP prepreg plies made of CYCOM 977-2 epoxy matrix and Toray T300 carbon fibers with quasi-isotropic [±452/0/904/0/90/02]s stacking sequence. The nominal thickness of the CFRP laminate was 5 mm, and a very thin fiberglass/epoxy ply, reinforced with 0°/90° fabric, was placed on the both sides of the CFRP laminate.

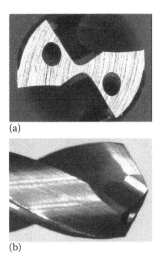

(a)

(b)

Figure 8.8 125° point angle twist drill: (a) front view and (b) side view.

The CFRP laminates were fabricated by hand lay-up, vacuum bag molding, and autoclave curing. The surface texture of the CFRP laminates on the bag side was rather irregular in comparison with that in the mould side, which displayed a very smooth surface texture.

Three traditional drilling tool types were utilized: a two-flute twist drill made of tungsten carbide (WC), with diameter 6.35 mm, featuring a 125°, 120°, or 115° point angle. The 125° point angle twist drill is shown in Figure 8.8a and b.

The aluminium alloy sheet was a 2024 T4 with 2.5 mm thickness. The titanium alloy sheet was a Ti6Al4V with 2.5 mm thickness. No cutting fluid was used during the drilling tests for all the cases under examination.

The process parameters, chosen according to aerospace industrial practice for manual drilling on assembly jigs, are reported in Table 8.6.

As it was decided to work without lubricant-coolant fluid, it was necessary to make a 3-mm-diameter partial prehole in the light alloy/CFRP stacks before final drilling, as reported in Figure 8.9. It is worth mentioning that, despite this provision, in the case of Ti/CFRP stacks, the maximum number of acceptable stack holes was never higher than 20 holes.

The trends of the thrust force and torque during drilling of light alloy/CFRP stacks are reported in Figures 8.10 and 8.11.

Four distinct phases can be identified in the graphs. The first phase (I) is related to the tool engagement in the workpiece and partial prehole drilling. The second phase (II) regards the drilling of the metal alloy. The third phase (III) concerns the drilling of the CFRP laminate. The fourth phase (IV) is related to the end of the drilling process and the tool exit from the work material.

Table 8.6 Process parameters of the light alloy/CFRP stack drilling campaign

Stack	Tool	rpm	a (mm/min)	f (mm/rev)	Holes Drilled (*n*)
Al/CFRP	125° point angle	1300	65	0,05	50
Al/CFRP	120° point angle	1300	65	0,05	50
Al/CFRP	115° point angle	1300	65	0,05	50
Ti/CFRP	125° point angle	1000	36	0,036	10
Ti/CFRP	120° point angle	1000	36	0,036	20
Ti/CFRP	115° point angle	1000	36	0,036	20

Source: Terracciano, R., *Caratterizzazione della Lavorazione di Foratura di Stacks Ibridi Composito Metallo*, MS thesis, Department of Materials and Production Engineering, University of Naples "Federico II," 2012.

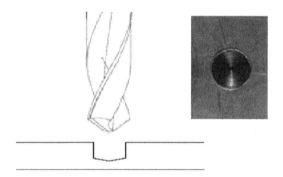

Figure 8.9 Partial prehole scheme.

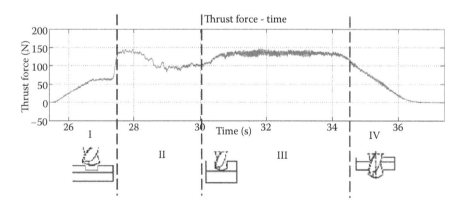

Figure 8.10 Thrust force trend during light alloy/CFRP stack drilling.

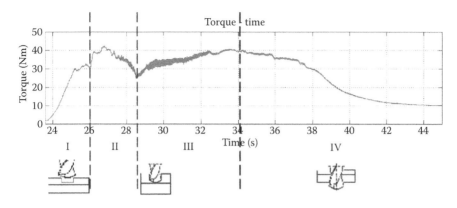

Figure 8.11 Torque trend during light alloy/CFRP drilling.

Also in the case of the torque, four phases can be identified in its trend, similarly to the case of the thrust force. The first phase (I) is related to the tool engagement in the workpiece and partial prehole drilling. The second phase (II) regards the drilling of the metal alloy. The third phase (III) concerns the drilling of the CFRP laminate. The fourth phase (IV) is related to the end of the drilling process and the tool exit from the work material.

In Figures 8.12 and 8.13, the trends of the thrust force and torque recorded during drilling of Al/CFRP stacks with a 125° point angle tool are reported.

The thrust force is practically the same for the diverse holes in the Al alloy sheet, while it varies from 140 to 250 N in the CFRP laminate going from hole 5 to hole 50. The trend of the thrust force in the Al alloy is largely conditioned by the presence of the pre-hole.

The torque shows a similar trend as for the thrust force: it does not vary in the Al alloy, while it varies from 35 to 45 Nm in the CFRP laminate going from hole 5 to hole 50.

In Figures 8.14 and 8.15, the trends of the thrust force and torque recorded during drilling of Al/CFRP stacks with a 120° point angle tool are reported.

The thrust force is practically the same for the diverse holes in the Al alloy sheet, while it varies from 50 to 280 N in the CFRP laminate going from hole 5 to hole 50. The trend of the thrust force in the Al alloy is largely conditioned by the presence of the prehole.

The torque shows a similar trend as for the thrust force: it does not vary in the Al alloy, while it varies from 10 to 55 Nm in the CFRP laminate going from hole 5 to hole 50.

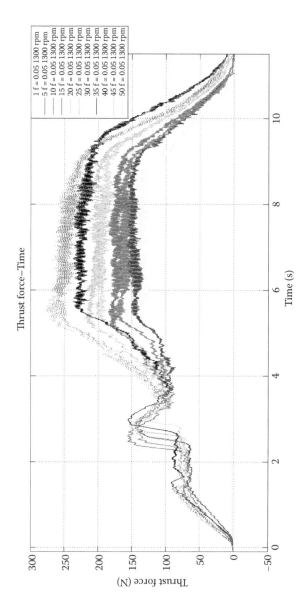

Figure 8.12 Al/CFRP stack drilling with 125° point angle tool: comparison of the thrust force recorded for every fifth hole in the drilling series.

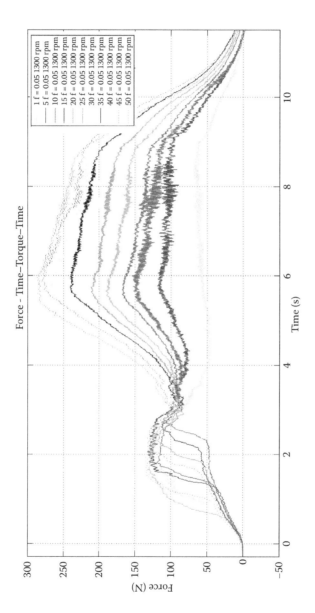

Figure 8.13 Al/CFRP stacks drilling with 125° point angle tool: comparison of the torque recorded for every fifth hole in the drilling series.

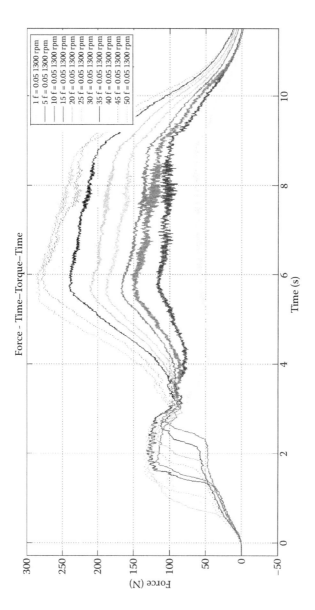

Figure 8.14 Al/CFRP stack drilling with 120° point angle tool: comparison of the thrust force recorded for every fifth hole in the drilling series.

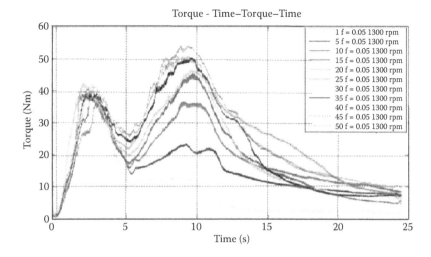

Figure 8.15 Al/CFRP stacks drilling with 120° point angle tool: comparison of the torque recorded for every fifth hole in the drilling series.

The thrust force and torque recorded in drilling of Al/CFRP, 115° point angle tool, are shown in Figures 8.16 and 8.17.

The thrust force varies from 100 to 150 N in the Al alloy sheet going from hole 5 to hole 50, while it varies from 50 to 280 N in the CFRP laminate going from hole 5 to hole 50. The trend of the thrust force in the Al alloy is largely conditioned by the presence of the prehole.

The torque shows a similar trend as for the thrust force: it does not vary in the Al alloy, while it varies from 20 to 50 Nm in the CFRP laminate going from hole 5 to hole 50.

In Ti/CFRP stack drilling, the thrust force is higher in the titanium sheet than in CFRP, contrary to what occurred in the case of drilling of Al/CFRP stacks.

The torque shows a similar trend as for the thrust force.

In the case of 125° point angle tool, it was possible to make only 10 holes, while for the 120° and 115° point angle tools, it was possible to make 20 holes.

In Figures 8.18 and 8.19, the trends of the thrust force and torque recorded during drilling of Ti/CFRP stacks with a 125° point angle tool are reported for all 10 holes made.

The thrust force is about 100 N in titanium sheet for the first hole and about 50 N for the CFRP. For the second hole, the thrust force in titanium is more than 250 N, more than double the previous value. This is due to the extremely high wear of the cutting lips of the tool. In the holes from 3 to 10, the thrust force varies from 250 N to 400 N, confirming that a very high wear of the tool has already occurred after the first hole.

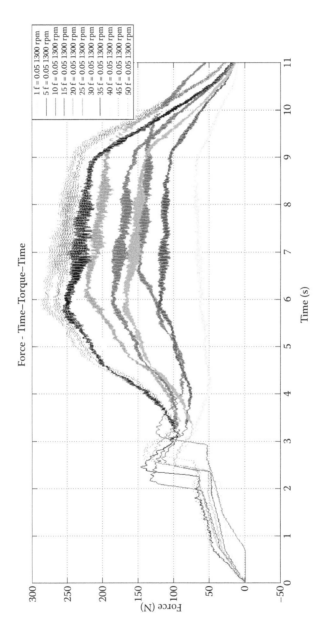

Figure 8.16 Al/CFRP stack drilling with 115° point angle tool: comparison of the thrust force recorded for every fifth hole in the drilling series.

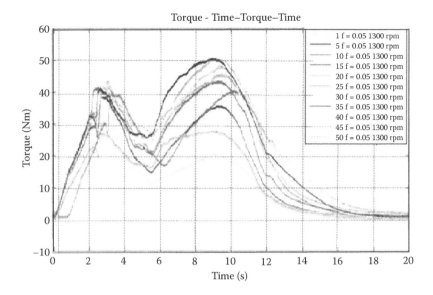

Figure 8.17 Al/CFRP stacks drilling with 115° point angle tool: comparison of the torque recorded for every fifth hole in the drilling series.

In Figure 8.20, the wear of the 125° point angle tool is reported. In the CFRP, the thrust force varies from 50 to 100 N. The thrust force in CFRP is influenced by the very high temperature of the tool.

The torque is higher in the titanium sheet than in CFRP. It shows a similar trend as for the thrust force: the difference between hole 1 and hole 2 is about 10 Nm, confirming that the wear of the tool already occurs at the first hole. The same trend is shown in the CFRP: the torque recorded for hole 2 is 10 Nm higher than the torque recorded for hole 1.

In Figures 8.21 and 8.22, the trends of the thrust force and torque recorded during drilling of Ti/CFRP stacks with a 120° point angle tool are reported.

The thrust force is about 150 N in titanium sheet for the first hole (not reported in the figure) and about 50 N for the CFRP. For the second hole, the thrust force in titanium is about 200 N. In this case, the difference between the first and second holes is quite less than for the 125° point angle tool, suggesting that wear progresses more smoothly.

The torque shows a similar trend as for the thrust force.

The thrust force and torque recorded in drilling of Ti/CFRP, 115° point angle tool, are shown in Figures 8.23 and 8.24. Also in this case, the wear progress smoothly and the torque shows a similar trend as for the thrust force.

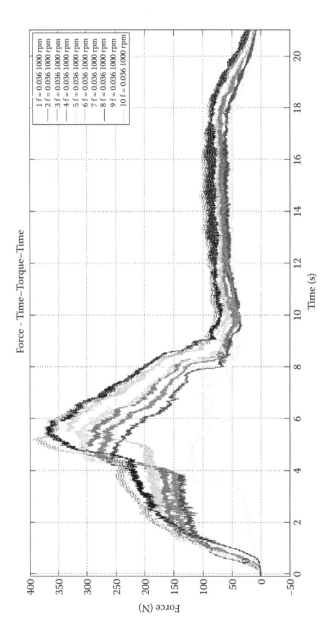

Figure 8.18 Ti/CFRP stack drilling with 125° point angle tool: comparison of the thrust force recorded for every hole in the drilling series.

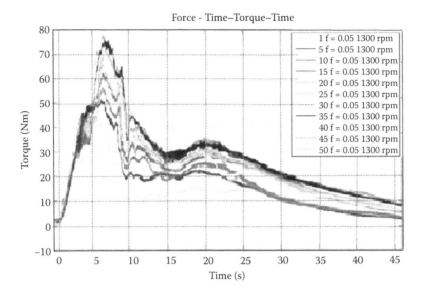

Figure 8.19 Ti/CFRP stacks drilling with 125° point angle tool: comparison of the torque recorded for every hole in the drilling series.

Figure 8.20 Ti/CFRP stack drilling: wear of 125° point angle tool.

The mean hole diameter decreases with increasing number of drilled holes for both Al/CFRP and Ti/CFRP stack drilling, as verified in the experimental light alloy/CFRP stack drilling campaign (Terracciano 2012).

The trend of the mean hole diameter measured after drilling the Al/CFRP stack with the 125°, 120°, and 115° point angle tool is reported in Figure 8.25a–c.

The trend of the mean hole diameter measured after drilling the Ti/CFRP stack with the with the 125°, 120°, and 115° point angle tool is reported in Figure 8.26a–c.

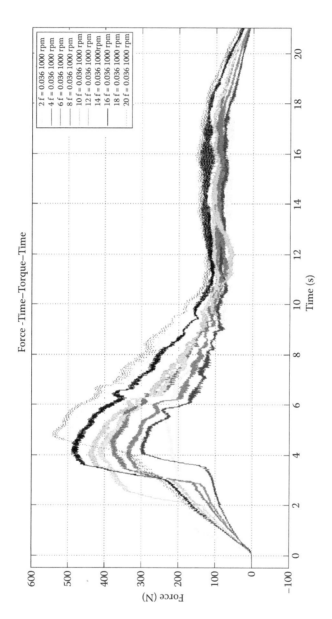

Figure 8.21 Ti/CFRP stack drilling with 120° point angle tool: comparison of the thrust force recorded for every second hole in the drilling series.

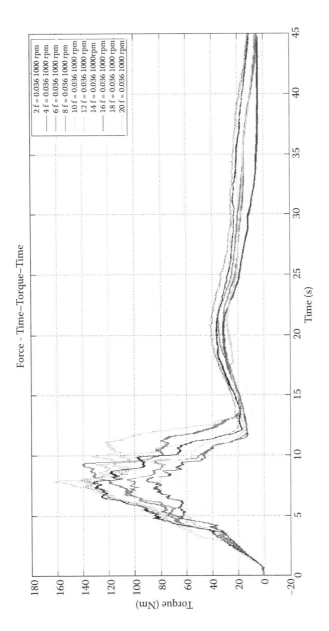

Figure 8.22 Ti/CFRP stacks drilling with 120° point angle tool: comparison of the torque recorded for every second hole in the drilling series.

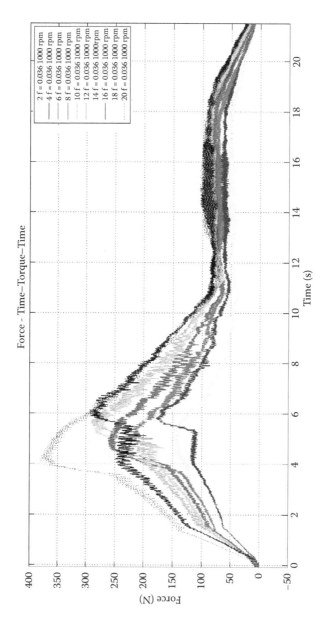

Figure 8.23 Ti/CFRP stack drilling with 115° point angle tool: comparison of the thrust force recorded for every second hole in the drilling series.

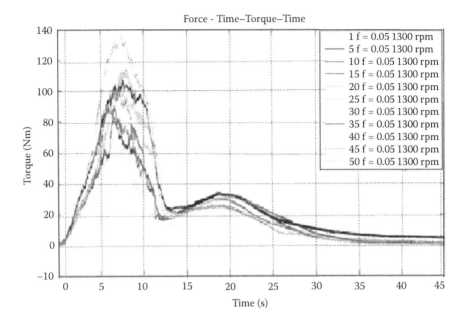

Figure 8.24 Ti/CFRP stacks drilling with 115° point angle tool: comparison of the torque recorded for every second hole in the drilling series.

8.7   Conclusions and future developments

In the manufacturing of light Alloy/CFRP hybrid stacks for airframe structures, the widespread technology for assembly is represented by conventional drilling and fastening, which, in 80% of the cases, is performed manually, especially for SMEs that supply subassemblies.

These technologies are highly labor intensive, and accordingly, a thrust toward automation is currently taking place in the aeronautical industry worldwide.

The automation of drilling processes applied to critical components such as those made of light alloy/CFRP stacks requires the development of optimized drilling operations and new tool geometries capable of decreasing the thrust force and torque during drilling in order to enhance the tool life and reduce the work material damage.

The utilization of online sensor monitoring, supported by advanced technologies of sensor fusion and machine learning allows for a deeper understanding of the physical phenomena involved in drilling of hybrid light alloy/CFRP stacks and prepares for the implementation of automated process based on robotic drilling integrating end-effectors with dedicated sensors and actuators.

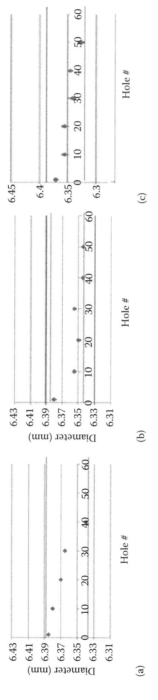

Figure 8.25 Mean hole diameter for Al/CFRP drilling with the (a) 125°, (b) 120°, and (c) 115° point angle tool.

Figure 8.26 Mean hole diameter for Ti/CFRP drilling with the (a) 125°, (b) 120°, and (c) 115° point angle tool.

In this chapter, the use of online sensor monitoring techniques during drilling of light alloy/CFRP hybrid stacks showed that tool geometry is a critical parameter. In fact, it was verified that even minor geometry variation, such as limited point angle variations, has a significant influence on the thrust force and torque maximum values and trends, determining a notable effect on tool life and work materials quality. The latter considerations are the more relevant in the case of Ti/CFRP hybrid stacks due to the low machinability characteristics of the Ti alloys.

- The maximum values of the thrust force vary with the point angle of the tool.
- The lowest maximum values of the thrust force were recorded for 120° point angle tool in the Al alloy.
- The lowest values of the maximum thrust force were recorded for 115° point angle tool in the Ti alloy.
- For the CFRP, the maximum thrust force values recorded are similar for the diverse point angles tested.
- The maximum values of the torque are unaffected by the point angle.
- The dimensional measurements of the mean hole diameters after drilling the Al/CFRP and the Ti/CFRP hybrid stacks suggest that the 120° point angle tool shows the best behavior in terms of lower reduction of the measured mean hole diameter with increasing drilled hole number.

It may be worth considering that the future technological developments for the robotic automation of hybrid stack drilling in the airframe manufacturing industry could benefit by and large from the implementation of the main key enabling technologies (KETs) of the Industry 4.0 paradigm. In particular, KETs such as intelligent sensors and actuators, machine learning, cloud manufacturing, cyber physical systems, internet of things, and big data analytics could make the realization of knowledge-based automation of drilling processes for the assembly of airframe fuselage feasible and effective, allowing for higher productivity, lower manufacturing costs, and reduced scrap rate through actual online and real-time process and tool condition monitoring. This automation can support the development of the smart factories of the future, where computer-driven systems monitor the physical processes and make decentralized decisions based on self-organization mechanisms without the need of human supervision.

References

Airbus, Airbus Global Market Forecast 2012–2031, 2012.

Bray JW, "Aluminum Mill and Engineered Wrought Products," in ASM Metals Handbook, Vol. II, *Properties and Selection: Non Ferrous Alloys and Special-Purpose Materials, 10th Edition*, 1990.

Buhl H, *Advanced Aerospace Materials*, Springer, Berlin, 1992.

Cayless RBC, "Alloy and Temper Designation Systems for Aluminum and Aluminum Alloys," in ASM Metals handbook, Vol. II, *Properties and Selection: Non Ferrous Alloys and Special-Purpose Materials, 10th edition,* 1990.

Cort A, "One Lean, Mean Airplane," *Assembly Magazine* (online), August 21, 2008, p. 1.

Heimbs S, Vogt D, Hartnack R, Schlattmann J, Maier M, "Numerical simulation of aircraft interior components under crash loads. *Int. J. Crashworthiness,* 13(5), 511–521, 2008. doi:10.1080/13588260802221203 2008.

Ho-Cheng H, Dharan CKH, "Delamination during drilling in composite laminates", *J. Eng. Ind.,* 112(3), 236–239, 1990.

Hoskin BC, Baker AA, *Composite Materials for Aircraft Structures,* American Institute of Aeronautics and Astronautics, New York, 1986.

ICAO, https://data.worldbank.org/indicator/IS.AIR.PSGR.

Immarigeon JP, Holt RT, Koul AK, Zhao L, Wallace W, Beddoes JC, "Lightweight materials for aircraft applications," *Mater. Charact.,* 35(1), 41–67, 1995.

Jerome P, "Composite materials in the Airbus A380," in: 13th International Conference on Composite Materials, ICCM-13, June 25–29, 2001, Beijing, China.

Kelly A, ed., *Concise Encyclopedia of Composite Materials,* Pergamon, Oxford, UK, 1994.

Owen D, Gardner S, Modrzejewski B, Fetty J, Karg K, "Improving wear and fretting characteristics with fiber reinforced aluminum liners," in: Proceedings of AHS 70th Annual Forum, Montréal, Québec, Canada, vol. 4, 2014, pp. 2597–2606.

Rooy EL, "Introduction to aluminum and aluminum alloys" in ASM Metals handbook, Vol. II, *Properties and Selection: Non Ferrous Alloys and Special-Purpose Materials", 10th Edition,* ASM International Handbook Committee, Metals Park, OH, 1995.

Smith F, *The Use of Composites in Aerospace: Past, Present and Future Challenged,* Avalon Consultancy Services Ltd., Newbury, UK, 2013.

Terracciano R, "Caratterizzazione della Lavorazione di Foratura di Stacks Ibridi Composito Metallo", MS Thesis, Department of Materials and Production Engineering, University of Naples "Federico II", 2012.

Weber A, "High-flying robots," *Assembly Magazine* (online), April 29, 2009, p. 2.

Welsch G, Boyer R, Collings EW, eds., *Materials Properties Handbook: Titanium Alloys,* ASM International, Metals Park, OH, 1993.

Ye L, Lu Y, Su Z, Meng G, "Functionalized composite structures for new generation airframes: A review," *Compos. Sci. Technol.,* 65(9), 1436–1446, 2005, ISSN 0266-3538.

Younossi O, Kennedy M, Graser J, *Military Airframe Costs: The Effects of Advanced Materials and Manufacturing Processes,* RAND Corporation, Santa Monica, CA, 2001.

Index

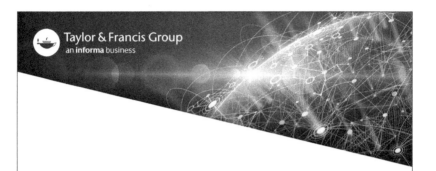